The Big One

Books by Jake Page and Charles Officer

A Fabulous Kingdom

Earth and You

The Great Dinosaur Extinction Controversy

Tales of the Earth

The BIG ONE

The Earthquake That Rocked Early America and Helped Create a Science

JAKE PAGE and
CHARLES OFFICER

Houghton Mifflin Company

BOSTON • NEW YORK

2004

For information about permission to reproduce selections from
this book, write to Permissions, Houghton Mifflin Company,
215 Park Avenue South, New York, New York 10003.

Visit our Web site: www.houghtonmifflinbooks.com.

ISBN-13: 978-0-618-34150-4   ISBN-10: 0-618-34150-1

Library of Congress Cataloging-in-Publication Data
Page, Jake.
The big one : the earthquake that rocked early America
and helped create a science / Jake Page and Charles Officer.
p.   cm.
Includes index.
ISBN 0-618-34150-1
1. Earthquakes—New Madrid Seismic Zone. I. Officer,
Charles B. II. Title.
QE535.2.U6P34 2004
551.22'09778'985—dc22   2004040536

Printed in the United States of America

Book design by Victoria Hartman

MP 10 9 8 7 6 5 4 3 2 1

We dedicate this book to the memory of
MYRON FULLER and OTTO NUTTLI,
who contributed so much to our understanding of the New Madrid earthquakes

ACKNOWLEDGMENTS

We wish to thank Arch C. Johnston and Buddy Schweig of the Center for Earthquake Research and Information at the University of Memphis for both their hospitality and their help. We are grateful to our indefatigable agent, Joe Regal of Regal Literary, for many things, notably for putting us in touch with our editor at Houghton Mifflin, Laura van Dam. A skilled science writer in her own right, she masterfully and diplomatically wielded both carrot and stick to help us shape this book.

The authors also are most grateful to their wives, Susanne and Trix, for the continuing patience and support in many forms that they bring to these ventures and without whom the world would be a far smaller place.

Contents

Introduction: Reelfoot's Folly

You stand at the end of a well-constructed boardwalk over the relatively opaque and windblown water of the lake. Cypress trees with their feathery green foliage stand sociably in the shaded water among the knobbly remains of their immediate ancestors. The bottoms of the tree trunks flare out just above the water line and are called knees. A few hundred yards out, a fisherman's dinghy bobs in the water. This is Reelfoot Lake, in the northwestern corner of Tennessee, and its dancing waters glisten in the sun. It does not look like earthquake country.

But Reelfoot Lake, some twenty miles long and up to six miles across, was created by the worst earthquakes ever to strike the lower forty-eight states of America.

To get here, you drive about a hundred miles north from Memphis, slowing down to pass through trim little towns strung along the mostly straight highway like beads. Here in these towns there is much manicuring of lawns, and Jesus abides, ubiquitous. It appears to be a serene and trusting existence in this part of Tennessee, and it just does not look like earthquake country.

A few miles closer than Reelfoot Lake to the huge muddy

Mississippi River, and beyond its western bank in the boot heel of Missouri, the land is flat—vastly, enormously flat—a place of huge agricultural fields broken here and there by long stands of trees, and on the horizon a long rise of land: a levee, designed to tame the river. Indeed the vast flat land appears to be thoroughly tamed, as most agricultural lands are. It is a serene landscape, if a monotonous one, and it simply does not look like what you would think was earthquake country.

In fact, this is part of the most seismically active region east of the Rocky Mountains and some fifteen hundred miles or so east of the San Andreas Fault, which, as everyone knows, is another way to spell "earthquake." Tremors, most of them too slight to be felt, occur here regularly, almost daily, and all you need to do if your goal in life is to feel an earthquake under your feet is to stand around this place for a few months. A Magnitude 4 could well be your reward.

Or instead of just standing here, you might try to find the Reelfoot Scarp. A scarp is an abrupt rise, like a cliff, and in this case the name would seem to be a bit of hyperbole on the part of the geologists. In most of its length the Reelfoot Scarp does not rise more than a foot or two above the surround, and at many places along its length it is simply not present for an untrained pilgrim. With superb directions and a nondigital odometer in your car, you might find it crossing State Route 78 a bit less than three miles north of Tiptonville, Tennessee. It is a tiny rise upholstered by corn plants and looking as mild as an old easy chair.

But whatever it was that caused the Reelfoot Scarp also created Reelfoot Lake. The scarp evidently rose high enough to cut off Reelfoot Creek, which, earlier, led into the Mississippi. The land sank and the waters built up, covering cypress forests that still can be seen, witchlike remnants with old dead branches swaying ever so slightly under the water. The locals will tell you that, sure, there was the Big One, but ponds, water, maybe even lakes always existed where Reelfoot Lake now lies. Probably true. This is a watery part of the world. After all, the Mississippi—originally the product of the rapid melting of the glaciers to the north some fifteen thousand years ago—has changed beds with the frequency of a call girl.

The locals, who seem a bit defiant, still also have a certain fondness for a group called the Night Riders who plied these lakeshores back around the turn of the twentieth century. There was a problem about whether the shores were to be the private preserve of a few landowners or open to the public. The Night Riders, adorned with masks made of wheat sacks, evidently terrorized the private landowners and, overcome with righteousness, also took to disciplining any local moral lapses—for example, whipping a young woman who wore clothes they judged to be too cheerful and bright and therefore provocative. After a few years of this sort of vigilantism, the state took over the lake and made it a state park, which it remains today, happily, and the Night Riders disbanded.

In addition to the not altogether denigrated antics of the Night Riders, there is an Indian legend that explains the odd name of the lake. It is said that the chief of a village nearby had the misfortune of being slightly crippled by a clubfoot. His gait was therefore hampered, and he was called Reelfoot. After several years of adulthood and chieftainship had passed, he still had not been able to attract a mate from among the young women of his village, so he set out for a village of Chickasaws, where he saw the girl of his dreams.

The girl was evidently no more pleased by Chief Reelfoot than any of the available women of his own village had been. So he went home and plotted to kidnap her. Overhearing this plan, the Great Spirit told Reelfoot that such an act was not appropriate, and were he to carry it out, there would be hell to pay.

Overwhelmed by desire, however, Reelfoot descended on the village, snatched up the beautiful maiden, and hauled her back toward his village, no doubt kicking and screaming. Not given to empty threats, the Great Spirit set the land to heaving and buckling, and beneath Reelfoot's feet it sank and filled with water, creating the lake named for him. To this day, it is averred, his remains and those of his inamorata reside in the mud of the lake bottom among the dead cypresses.

That this explanatory tale may have arisen sometime after 1812 is suggested by the fact that it was in February of that year that the last of three rapid-fire great earthquakes devastated thousands of square miles of the regional landscape, heaving it around as if, in truth, the Great Spirit was annoyed. In the process, the trio of quakes created some ten new lakes, of which one was Reelfoot. These earthquakes were, in combination and in fact, the Big One. They were the most widely felt quakes ever known. They occurred in a place where, as one geologist has said, there is no reason to have an earthquake. And they remain an enigma.

This book is an effort to describe these terrible upheavals and, by way of looking at the history and progress of the earthquake sciences worldwide, to explain what we know about the nature and causes of earthquakes in general and, in particular, the oddball assault on New Madrid, the little river town in Missouri that, however willingly or unwillingly, lent its name to these events.

It is a story, of course, that just might have to be told again when Mother Earth or the Great Spirit or whatever forces are at work underneath this seemingly placid region have another grand mal seizure.

PART ONE

THE NEW MADRID QUAKES

Then the earth shook and trembled; the foundations
also of the hills moved and were shaken, because he was
wroth.

— Psalms 18:7

Do not disturb yourselves at the quaking . . . nor do
you imagine that this earthquake is a sign of another
calamity; for such affections of the elements are to the
course of nature, nor does it import anything further to
men than what mischief it does immediately of itself.

— King Herod, 31 B.C.

The World Gone Mad

Day One

A TREMENDOUS ROAR, the ground writhed in convulsions. People thrown from their beds, screaming, running out into the icy cold, the waters of the Mississippi heaving up and crashing down.

Sharp explosions — trees splitting, thousands of trees, snapping like rifle fire, acres of riverbanks plunged thunderously into the river, crushing under tons of mud uncounted flatboats moored for the night.

The earth heaved up, collapsing, ancient waterlogged trees ejected from below the river's waters, thrown high.

A blanket of malodorous fog thickened the air, obscuring the stars; fissures and cracks opened in the earth underfoot, then slammed shut, sending geysers of liquid sand, black soot, sulfurous gases exploding high into the air; chimneys collapsed; cataracts gushed in the river, turning onetime channels into violent rapids.

River islets turned to quicksand. Flashes of eerie light. Lakes vanished into the earth, becoming dry land. The very gates of Hell opened.

Around two o'clock on the morning of December 16, 1811, a profound shuddering of the earth began. The mighty Mississippi River ran backward? Had the moon struck the earth? A Shawnee prophet called Shooting Star had said the Great Spirit was enraged. Subterranean fires threw down the arches or vaults of the earth, abyssal waters combusted, fermented . . . or electric fluid pervaded the bowels of the earth, driven by volcanic impulses from as far off as the Andes, heaving the ground upward in coruscations and explosions. An inexplicable catastrophe, the very world itself gone mad.[1]

Near the pioneer town of New Madrid, 140 miles below the mouth of the Ohio River, the ground shook incessantly, "like the flesh of a beef just killed," as one resident later put it. Chimneys fell and people took up residence in temporary campgrounds outside the town.

In the river town of Big Prairie a hundred or so miles to the south, everyone managed to escape, but long after dawn's light came, another quake struck and the town itself, built on sandy sediments, sank without a trace as the ground below it was saturated with water.

The land was in motion, undulating, like an ocean.

In Livingston County, Kentucky, homesteader George Crist and his family were knocked out of bed: "The roar I thought would leave us deaf if we lived. When you could hear, all you could hear was screams from people and animals. . . . I thought the shaking and the loud roaring sound would never stop . . . we was all banged up and some of us knocked out for awhile and blood was everywhere."

After the first great shock, more followed before dawn —

shakes they were called — some people counted seven more. A noise like distant thunder, and the ground heaving upward, a thick haze, nausea, giddiness. The earth endlessly undulating as if God was shaking out a cosmic bedsheet.

Did the comet strike a California mountain? Are there three righteous persons in Natchez? Or like Sodom and Gomorrah, is Natchez too destroyed? Rumors were rife. Is the city of New Orleans sunk? Trees rose up from the river bottom. Trees sank in new lakes. In Charleston, South Carolina, the bells of St. Philip's Church rang. The parade grounds in Georgetown, South Carolina, sank two inches.

Hardest hit, perhaps, on that first day was the town of Little Prairie, the small and unprepossessing river settlement of some hundred or so people. Thrown from their beds into a murky black night redolent with foul odors, unnatural lights flashing from the trembling ground, the townspeople saw cracks in the ground everywhere, crevasses opening with hissing sounds, liquid sand and warm water rising, mud . . . nowhere to step. Shocks came through the night, with a huge one late in the morning, striking — as one survivor later put it — as if from right below, "with a perpendicular bouncing that caused the earth to open in many places — some eight and ten feet wide." Waterspouts erupted, eight feet into the air, and the town sank as a dark liquid rose.

In the morning, old George Rodell, the village elder, watched his mill tip over as the riverbank gave way and his house collapsed. Fifteen minutes after the eleventh shock, the water rose waist deep for as far as one could see. Moses-like, Rodell led the townspeople through the water, none of them knowing if or when they would step into a crevasse beneath the muddy waters and disappear — or step into quicksand and disappear. Coyotes, snakes, and other wildlife struggled through the flood among them. Spouts of water and black dust erupted around them into the air as if they were under cannon fire. Through the winter cold and waist-deep water they trudged, holding small children and a few possessions above the water, in all a terror-filled trek of eight miles to high ground. From there, they eventually made their way to New Madrid twenty-six miles away. They arrived eight days later on Christmas Eve, finding that hoped-for haven destroyed, its people in hardly better shape than they themselves.

Three months later, a merchant named James McBride passed through these parts. He found little left of Little Prairie but a few coffins projecting out of the riverbank, part of a cemetery eaten away by the river along with most of the town site. Instead, the ground was cracked and pockmarked with large round holes surrounded by two-foot-high ridges of sand and a black coal-like material. "All nature appeared in ruins," McBride wrote. Overcome by a sense of dread, he boarded his boat and was hastily on his way.

Close Observers

It would be some time before the extent of the damage from the quake of December 16 was known, or the enormous range of the shocks. As reports arrived in the offices of newspapers over the next days and weeks, it became clear that the area over which shocks were felt was unprecedented: a million square miles. Significant damage occurred in an area about the size of Texas and minor damage over a region twice that size. (By comparison, damage caused by the storied San Francisco earthquake of 1906 occurred over an area smaller than that of Delaware.)

The white settlers in the region of significant damage probably numbered no more than a few thousand: this was still a lonely and remote frontier for people of European descent. One could travel the Mississippi and many of its tributaries, floating for a hundred miles or more between settlements with little but the occasional smoke rising from an Indian town or encampment. How many people were killed in the shocks of December has been estimated since at a maximum of fifteen hundred, with most of those being Indians, but no one will ever know the exact number of lives taken. In any event, it was predictably very few for so great a quake. What might seem strange to anyone but a statistician is that a surprising number of careful and close observers of the quakes and their effects were in the neighborhood when the major quakes occurred. Among them were naturalists (including the artist and naturalist John James Audubon), a writer, an engineer, and a former sea captain who was at the time in the diplomatic service.

Audubon was riding in what he called the barrens of Kentucky when he saw the sky go dark in the west. A mile later, he heard a distant rumbling, perhaps of a tornado, and he spurred his horse to gallop, but it refused to do so. Instead, his mount "placed one foot after another on the ground, with as much precaution as if walking on a smooth sheet of ice. I thought he had suddenly foundered." Audubon was about to dismount when the horse began "a-groaning piteously, hung his head, spread out his legs, as if to keep himself from falling, and stood stock still, continuing to groan." Again Audubon decided to dismount, but "at that instant all the shrubs and trees began to move from their very roots. The ground rose and fell in successive furrows, like the ruffled waters of a lake."

Another naturalist found himself right in the middle of the quake's destruction, being rudely awakened at about two o'clock on December 16 while sleeping aboard a flatboat laden with fifteen tons of lead. This was John Bradbury, a Scot trained at the University of Edinburgh and on assignment from the Botanical Society of Liverpool to assay the plant life of western America. He had arrived earlier in St. Louis with a letter of introduction from Thomas Jefferson, and had traveled some two thousand miles up the Missouri River, as far as the villages of the Mandan Indians. This was largely unexplored territory, and the fur traders who accompanied Bradbury reined him in from time to time for his own safety among these little-known Indians. (Bradbury grumbled in print about this curtailment of his freedom when he wrote up his trip in an influential volume, *Travels in the Interior of America, in the Years 1809, 1810, and 1811*.) Now trying to reach New Orleans, the ever-confident and adventurous botanist had agreed to take charge of a boat carrying lead for sale in that port city, the crew consisting of four French oarsmen and a skilled French steersman, or *patron*. They left the town of New Madrid on December 14, and arrived a hundred miles downstream at a place called Chickasaw Bluffs just above a particularly dangerous channel called the Devil's Race Ground on the fifteenth. Rather than attempt this channel in the crepuscular light of oncoming night, they moored to the sloping bank of a small island. This saved their lives.

When the shock struck, it was accompanied by a "most tremendous noise," and the boat lurched upward, in evident danger of capsizing. While the crew cried out in something of a panic, Bradbury went outside and saw the river tossed as by a storm, covered with foam and drifting trees and branches. While "the noise was inconceivably loud and terrific, I could distinctly hear the crash of falling trees, and the screaming of the wild fowl on the river . . . all nature was in a state of dissolution."

Bradbury went ashore and found the bank split off by a chasm some eighty feet long and four feet wide, but it still held, while elsewhere the vertical banks of the river were caving in with tremendous crashes. He persuaded the crew to stay on board, the boat being safer than the ground. Several canoes floated past them, empty of people, "melancholy proof that some of the boats we passed the preceding day had perished." The shocks continued, smaller ones but a few of them large enough to nearly throw the men off their feet, in all some twenty-seven through the night. Counting them, Bradbury also noted that "the sound which was heard at the time of every shock, always preceded it at least a second, and that it uniformly came from the same point, and went off in an opposite direction." It came, he continued, "from a little northward of east, and proceeded to the westward." Such precise observations would be of considerable value to earthquake scientists almost two centuries later.

As the day wore on, Bradbury and crew made their way through the Devil's Race Ground and on to the south amid continuing shocks, the crashing of trees, and the screaming of wildfowl. Near the Lower Chickasaw Bluffs, Bradbury put ashore to visit briefly with some people huddled prayerfully on the river's edge. Most of the people in their settlement had fled the scene. One of the men attributed the quake to a comet that had been plying the sky, predicting that the end of the world was likely. "Finding him confident in his hypothesis," Bradbury wrote, "and myself unable to refute it, I did not dispute the point." Later on the way south to Natchez, as the shocks became fewer and farther apart, Bradbury heard from another traveler that the shocks had been extremely violent at New Madrid: "The greatest part of the houses had been rendered uninhabitable, although, being constructed of timber, and framed together, they were better calculated to withstand the shocks than buildings of bricks and stone."

One of the more bizarre, or one might say dramatic, side effects of the quakes' effects on the river concerned a family by the name of Sarpy. Some two hundred miles north of Natchez lay an island that was known by several names, among them Crow's Nest. On the night of December 15, a Captain Sarpy from St. Louis and his family tied up on the northern end of the island preparatory to spending the night. Sarpy left his boat to look around and came upon a group of river pirates, who, it seems, were expecting Sarpy to show up and planned to rob him. So Captain Sarpy sneaked back to his boat and dropped downriver to the next island, where he tied up for the night. The Sarpys awoke to see that the island known as Crow's Nest had disappeared under the water, and along with it, the pirates.

The river also provided a daunting test for the maiden voyage of the first steam-driven vessel to ply the Mississippi, the *New Orleans,* which was also the first steam-driven vessel built in the West and entirely from western timber and iron. The sternwheeler was 116 feet from bow to stern, 20 feet across at its widest, and it displaced 410 tons. Setting out on the Ohio River (only three years after the first steamboat had plied the Hudson River back east), its great clouds of steam and belching sparks caused a considerable stir, even fear. In December, the *New Orleans* was north of New Madrid and felt the shock of the December 16 quake while moored to an island. But subsequent shocks were masked by the sound and vibrations of the thirty-four-piston engines that drove the boat through the water: the crew could *see* the effects, such as whole forests waving without benefit of a breeze. When she passed New Madrid and other sites where people were crying out in dismay, she took on no passengers, in part because many were more afraid of the boat than the quakes. Eventually, with the normal navigational markers disrupted and the boat keeping to those places where the current was strongest, she made it to Natchez on January 2, two thousand miles from where she had started. There she was boarded by none other than the Scot John Bradbury, who pronounced her a "very handsome vessel."

In the first week of January 1812, New Madrid was visited by an extraordinarily colorful adventurer with as checkered a career behind him and in store as a historical novelist could ever need.

This was Louis Bringier, the son of a fabulously successful plantation owner in Louisiana's St. James Parish who also owned a fine townhouse in New Orleans. Young Louis had received an excellent education — probably back in France, where wealthy colonists usually sent their sons — and probably a scientific education of some sort. Back home, he had been awarded a considerable Spanish land grant in Louisiana by a departing Spanish governor, but by 1810 he had managed to run up some huge gambling debts. Evidently in some humiliation, he left town and went native, living among the Cherokees and Osages for two years, letting on later that he had become a "chief."

(Bringier, to leap ahead, later returned to New Orleans and became an important member of society, the city surveyor and then surveyor-general for the state of Louisiana. But before reachieving such high status, he had participated in the Battle of New Orleans, during which time, no doubt, he met the infamous pirate Jean Lafitte, who had terrorized the port of New Orleans for several years. Lafitte, it is not widely known, supplied much of the armament that made Andrew Jackson's defeat of the British at New Orleans possible, and he was pardoned along with his men by a grateful President James Madison. Then Lafitte and Bringier were sent on what amounted to a spy mission into the Arkansas Territory, posing as leaders of a mining expedition. At some point, as well, though the timing is not clear, Bringier also went to Mexico and supposedly made a fortune in silver mines but lost it when he got involved in a revolution there.)

In any event, it is certain that Bringier — during his "native" stage — was in New Madrid in early January 1812 and experienced a large shock while there. His description, along with valuable observations of the Indians of the region, appeared in 1821 in the leading scientific journal of the period, the *American Journal of Science and the Arts*. He gave the date of January 6 for the date of this shock, but it probably happened on January 7, based on records found later. The particular value of his report is that he concentrated on the ground damage and the huge amounts of liquid thrown up fifteen feet into the air, along with carbon dust that "fell in a black shower, mixed with the sand." At the same time, he wrote, "the roaring and whistling produced by the impetuosity of the air escaping from its confinement, seemed to increase the horrible disorder of the trees which everywhere encountered each other, being blown up, cracking and splitting, and falling by the thousands at a time. In the meantime, the surface was sinking, and a black liquid was rising up to the belly of my horse, who stood motionless, struck with terror."

He remarked as well on the blowholes that covered the country like "so many craters of volcanoes, surrounded with a ring of carbonized wood and sand." He returned to New Madrid later, remarking that the countryside that was formerly woodlands interspersed with little prairies and meadows was now covered with ponds and hills of sand.

⌒

On December 25, only nine days after the first great shock, another traveler wrote his observations to the editor of the *New York Evening Post*. William Leigh Pierce, the son of a Georgian who was one of the lesser-known delegates to the Constitutional Convention, was on a river trip from Pittsburgh to New Orleans on December 16. His account was among the earliest to alert the larger American public to the actual events that had shaken much of the country, giving rise to alarums and rumors of all kinds, including the belief in some quarters that a volcano had arisen on December 16 in North Carolina.

Like Bradbury, Pierce had moored his craft to a small island south of New Madrid and he experienced much the same by way of nighttime shocks, hearing "tremendous and uninterrupted explosions, resembling a discharge of artillery. . . . Wherever the veins of the earthquake ran, there was a volcanic discharge of combustible matter to a great height, as incessant rumbling was heard below." The great spouts of water were especially alarming: "Large trees, which had lain for ages at the bottom of the river, were shot up in the thousands of instances, some with their roots uppermost and their tops planted; others were hurled into the air; many again were only loosened, and floated upon the surface. Never was a scene more replete with terrific threatening of death."

During the day, the boat moved downriver, tying up to another island, one of those that had not disappeared. There the earthquakes "had rent the ground in large and numerous gaps; vast quantities of burnt wood in every stage of alteration, from its primitive nature to stove coal, had been spread over the ground . . . hideous caverns yawned on every side, and the earth's bowels appeared to have felt the tremendous force of the shocks which had thus riven the surface." The great spouts they had been so alarmed by, Pierce found, "were generally on the beach; and have left large circular holes in the sand, formed much like a funnel. For a great distance around the orifice, vast quantities of coal have been scattered, many pieces weighing from 15 to 20 lbs. were discharged 160 measured paces. These holes were of various dimensions; one of them I observed most particularly, it was 16 feet in perpendicular depth, and 63 feet in circumference at the mouth."

Elsewhere in his report, Pierce noted that the land near Fort Pickering, 242 miles below the mouth of the Ohio, was "strong and high," and others would report that people and structures on high ground were largely undisturbed or did not feel the shocks at all while just below in valleys the earth undulated like waves of water. Also, Pierce took note of what he took to be "traces of prior eruptions, all of which seem to corroborate an opinion that the river was formed by some great earthquake — [I have no doubt] that there are at the bottom of the river strata upon strata of volcanic matter." Pierce included in his report a table of the shocks he experienced — in all eighty-nine from the first one on December 16 until they evidently stopped on December 23 — and he made note of the fact that the river's current was calculated to be six knots at dawn on December 16, faster by at least one knot than the river typically flowed at full flood.

"The continuance of this earthquake," Pierce wrote, "must render it conspicuous in the pages of the Historians, as one of the longest that has ever occurred." The observant young man, writing on Christmas Day near where Big Prairie had once stood, quite literally did not know the half of it.

Continuance

If the New Madrid quake had turned out to be no more than the huge December 16 shock, followed by a long series of aftershocks over the next week or so, it would have been considered a fairly normal, if immense, earthquake. Most likely it would have been largely forgotten amid the grand ambitious hurly-burly of becoming a nation, and not of any special puzzlement to earthquake scientists in the twentieth century and beyond. Most earthquakes, after all, consist of a large shock and a number of aftershocks diminishing in potency. But in the New Madrid area, there were two more mammoth shocks — one on January 23, 1812, and yet another on February 7. Compared to these two, the shock experienced by Louis Bringier on January 7 was not major. Smaller shocks occurred throughout, and continued on at a rate of better than one a day for months. Indeed, the ground around New Madrid continued to shake — so much so that as the years went by, permanent residents of the area paid little attention to such common instances of *terra tremula*.

On the other hand, one can hardly imagine the horror with

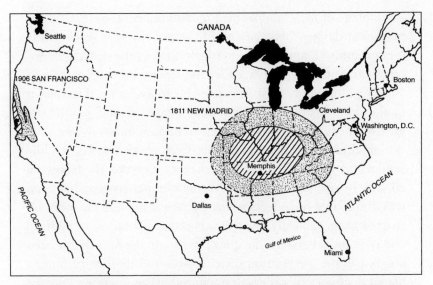

Figure 1. A comparison of the New Madrid earthquakes of 1811–1812 and the San Francisco earthquake of 1906. The hatched area shows the zone of the most severe shaking and damage. The dotted area shows the zone where damage was less severe but the ground shaking was felt by all. Adapted from W. W. Hays, U.S. Geological Survey Professional Paper 1240-B, 1981.

which the locals must have confronted a second enormous quake about a month after the first, and *then* a third, shattering their world yet again only some two weeks later. Those whose devotion to the letter of the Ten Commandments may have lapsed (being so far from the peer pressure found in centers of "civilization") surely saw this as something of an artillery attack from a wrathful God. According to Timothy Flint, a Presbyterian missionary from Connecticut who plied the frontier for some ten years and wrote about it in 1826, the people of New Madrid "had been noted for their profligacy and impiety." But with the onset of the shocks, Catholics and Protestants all became one religion and "all betook themselves to the voice of prayer." Even the town's cattle joined up in their bovine way, crowding "about the assemblage of men, and seemed to demand protection, or community of danger." The victims of the quakes were surely ready for redemption, and in the months following the quakes, ministers swarmed into the area, collecting repentant sheep in flocks.

Long afterward, scientists were able to assess the three major quakes in terms of their magnitude, a capacity and a system of measurement to which we will return in due course. By one such system, the December 16 quake has been calculated to be 8.2; the January 23 quake 8.1; and the February 7 quake, the largest of all, 8.3. By way of comparison, the San Francisco quake of 1906 was 7.6. While the effects of the 1906 quake were far more restricted geographically, the site was obviously far more built-up and populous. Between the quake itself and the fires that started afterward, the devastation of structures and the loss of human life were greater by orders of magnitude than anything inflicted by all three of the New Madrid quakes. But one of many questions that would await answers from twentieth-century scientists was why, since they all were of similar magnitude — which is to say that a similar amount of energy was explosively released — the effects of the New Madrid quakes reached so much farther across the land than the quake in northern California. For it would become clear after the February 7 quake that it had been felt as far off as Montreal, and virtually everywhere in what was then the inhabited United States and its territories, all the way to the Atlantic coast in Florida.

Few firsthand accounts exist of the second quake. A brief and sad one came from George Crist, who had resolved after the first terrifying great shock of December to take his family — wife Betsy and seven children — and move eastward. But when the January 23 quake rocked the earth, they were still in the state. In his journal, Crist wrote, "The earth quake or what ever it is come again today. It was as bad or worse than the one in December. We lost our Amandy Jane in this one — a log fell on her. We will bury her upon the hill under a clump of trees where Betsy's Ma and Pa is buried. A lot of people thinks that the devil has come here. Some thinks this is the beginning of the world coming to an end."

The reason for the paucity of accounts, according to historian James Lal Penick, Jr., was the icy cold of deep winter on the Mississippi. The northern part of the globe was still in what has been called the Little Ice Age, which had begun nearly four centuries before. Temperatures in the Ohio and Mississippi valleys were considerably lower then. It was not uncommon for the Mississippi itself to freeze over. Throughout the New Madrid area, most of the streams and tributaries to the mighty river were still frozen in late January when the quake hit. Boat traffic had ceased altogether on the river, and it was typically boatmen who brought news to the world of the quakes of December — and later of February.

Another account concerns a Baptist minister, Elder Wilson Thompson, who was making his icy way back home north of Cape Girardeau from New Madrid when this second great shock struck. His party's horses staggered and nearly fell, people near the river were screaming as their houses fell in, and trees snapped from "the great agitation of the earth." The shock (and those following) were preceded by the sound of distant thunder and of a huge wind — but "not a breeze of wind could be perceived." After the major shock, the party went on its way, noting that while it remained cold, it was warmer than before. The January shock would seem to have been pretty much a reprise of the one in December.

A little more than a week later, on February 4, the people who remained in New Madrid felt the earth twitch. It was, according to resident Eliza Bryan, writing several years later, "in continual agitation, visibly waving as a gentle sea." This nauseating heaving continued through days and nights until, at 3:15 in the morning of February 7, another major earthquake struck. It would later be referred to as "the hard shock." A vast riverine wave of unknown height swept over what was left of the town, swamping boats, breaking off cottonwood trees, and killing an unknown number of residents.

Some two thousand refugees from New Madrid and the surround all headed for some high ground forty miles to the north and seven miles west of the river. Only minutes after the hard shock struck in New Madrid, church bells rang in Pennsylvania and south to South Carolina. Windows rattled for two minutes in Montreal, more than a thousand miles to the north and east. A newsman in Cape Girardeau, a bit north of New Madrid, said the shock lasted almost fifteen minutes — an eternity for an earthquake. In St. Louis as well as other cities in the region like Cincinnati and Louisville, chimneys and brick houses collapsed. Soldiers far to the north and west in Fort Osage on the Missouri River were sure they were under attack by Indians. In Washington, D.C., people thought their houses might be being burgled.

The Mississippi had, only in recent days, become navigable again, and a Captain Mathias Reed reported encountering two sets of six-foot falls that had been suddenly created in the river near New Madrid. In the days afterward, Reed and his crew observed almost thirty boats capsize in these falls, their crews drowning. Convinced that the earthquakes were by no means over, Reed sold his boat and walked overland to St. Louis. He later wrote his recollections for the *Pennsylvania Gazette*, telling the harrowing tale of a man marooned on an island just north of New Madrid and holding on to a tree for support. Suddenly, the earth collapsed under him. "The fissure was so deep as to put it out of his power to get out of that place. He made his way along the fissure until a sloping slide offered him an opportunity for crawling out."

In another terrifying event, a highly successful merchant,

Vincent Nolte, was aboard one of his fleet of twenty boats a few miles south of the falls that had erupted in the river below New Madrid. Nolte determined he would be safer if his boat remained moored, but the other nineteen cast off into the dreadful turbulence of the river, now filled with trees, stumps, other craft, and many dead, and disappeared for good in the night and the torrent.

In New Madrid itself, the remaining houses were all destroyed. The high banks on which what was left of the town sat now fell some twenty feet down into the river. The rest of the site washed away in the spring floods of that year. (The same would be true for countless Indian villages along the area's rivers.) What was the original New Madrid, in Missouri, is now a sandbar in the river and woodland in a tiny fragment of the state of Kentucky that exists in the upper bulge of the river's oxbow. While no one can pinpoint with any great precision the actual epicenter of any of the three large quakes, the third quake's epicenter had clearly moved north and east and was only miles away from New Madrid.

Sunk Country

Even before the quakes of 1811 and 1812, the area that now comprises southeastern Missouri and northeastern Arkansas was a strange landscape. It consisted mostly of bottomlands, much of which were covered by standing water some or all of the time. Just before the quakes, an unusually frank booster of the area, Amos Stoddard, wrote that the land between New Madrid and Cape Girardeau to the north "is more insalubrious than any other part of Upper Louisiana." He went on to point out that "mephitic exhalations from the swamps and lowlands must necessarily poison the air; they produce intermittents, and some bilious fevers, though they have never been considered as very ma-

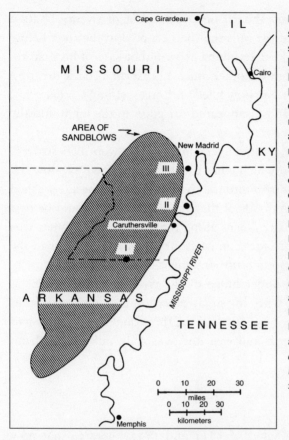

Figure 2. Map showing the area of sandblows, the locations of New Madrid and Little Prairie — known as Caruthersville on modern maps — and in roman numerals the locations, or epicenters, of the three main quakes. Adapted from both Myron L. Fuller, "The New Madrid Earthquake," published as the *U.S. Geological Survey Bulletin*, 494, 1912, and Otto Nuttli, "The Mississippi Valley Earthquakes of 1811 and 1812," *Seismological Society of America Bulletin*, 63, pp. 227–248, 1973.

lignant." He added that most deaths were the result of old age and accidents. With these unpleasant attributes out of the way, Stoddard then emphasized the large amount of forest, prairie, and good agricultural land in the region.

It was not, then, altogether "normal" country that was rearranged by the quakes. Indeed, many later accounts of the area suggested that the land had already suffered from at least one earlier earthquake. How else to account for the sunken lands of the region, the innumerable low ridges that separated them?

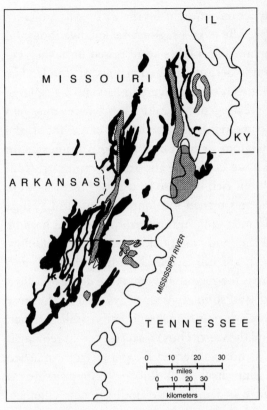

Figure 3. Map showing sunken lands (solid) and uplifts (dotted) caused by the New Madrid quakes. Derived from Myron L. Fuller, "The New Madrid Earthquake," *U.S. Geological Survey Bulletin,* 494, 1912.

With the oratorical flourishes appropriate to a politician, Missouri senator Lewis F. Linn wrote to his committee in 1836 that before the 1811–1812 quakes, "the greater portion of this gloomy region, annually covered by water, and at all seasons by a heavy growth of timber, and thick canebreaks, closely interwoven with plants of the convolvulus order, was *once* high ground." Many of the swampy areas, it was widely thought, had formerly been the actual course of the Mississippi and its tributaries, which had moved as a result of earlier earthquakes. (These rawboned suggestions would, in fact, be proven to be the case much later.)

Even so, the destruction wrought by the 1811–1812 quakes was profound — especially over a region of some five thousand square miles. Swamplands and lakes were raised until they became dry land. Streams became instead strings of ponds. The quakes threw up dams that caused rivers to become lakes; large areas of land sank, creating ponds and lakes where none had been before. In all, some ten good-sized lakes came about, the largest of which is Reelfoot Lake in northwestern Tennessee. In many of these lakes, one could see the forests rising up from the lake bottom. (All but two of these lacustrine creations were subsequently drained and turned into agricultural land.) Huge areas of agricultural lands sank in the earthquakes and became swamps. All around the land was pockmarked with sandblow craters, many of which are still visible. Fissures, crevasses, and fault lines still crisscross the region. Several large domes reached up out of the ground, and so many waterways west of the river were lost that in mid-century it was largely impossible to reach the river by water from the west. Those waterways that remained whole were as often as not clogged with logjams from smashed forests. In the Mississippi itself, well-known channels were no more, and navigation was a nightmare until the new channels and new landmarks became familiar to boatmen.

What was already known as sunk country, then, was further sunken, further battered, and reconfigured. Four out of the seven Euro-American settlements ranging along the river from Cape Girardeau to Natchez disappeared, along with one fort and two islands. One of the towns sent into oblivion by the quakes was a mill town named Spanish Mill, on the Little River, a tributary to the St. Francis River, which joins the Mississippi in Arkansas. The immediate damage to the town was reparable, but after the quake the town itself was surrounded by new swampland, and the Little River lost enough of its flow to make it too weak to run a mill.

In the years following, a number of people would continue to monitor the trembling earth and attempt to assess the upheavals' actual reach, as well as determine what could possibly have caused such a calamitous event. Answers to this last item would be slowest in coming, calling for the development of an entire new branch of science, and it has proved elusive still. Meanwhile, a more prosaic question presents itself: What were people doing in these sunken lands in the first place?

2

Dreams, Omens, and War

THE TOWN OF NEW MADRID, titular center of the New Madrid quakes, had been advertised in 1788 as the central feature of a grand colonization scheme. The scheme was born of both an admirable idealism and a less admirable expediency. Almost immediately, and well before it was shattered by the earthquakes to which it lent its name, New Madrid became what architectural historian John W. Reps called "one of the most interesting, and perhaps the most futile, episodes in American town planning."[1] The town plan was the brainchild of Colonel George Morgan, visionary, land speculator, minor player in the American Revolutionary War, and a man whose reach was more often than not greater than his grasp.

In planning New Madrid, Morgan was in cahoots with the Spanish minister to the brand-new United States, Don Diego de Gardoqui, whose chief role was to help see to it that this brash young nation did not extend itself to the west beyond the waters of the Mississippi River, where Spain had nominally held sway for some two decades. The Spanish themselves were confined chiefly to the port city of present-day New Orleans, though their

"Louisiana" (which they were to cede to the French in 1800), extended far to the north and west. The Spanish imperial presence constituted an awkward and very real threat to the upstart American nation in its early decades of existence. At the time, the government of the United States was essentially provisional and largely powerless, awaiting its true birth under the U.S. Constitution. The Scots-Irish immigrants who had swept into the lands across the Appalachians and east of the Mississippi River had little faith that the new government, even once duly constituted, would be able to help them much. Indian tribes like the Shawnees resented their presence, and violence broke out often, leading to spiraling instances of revenge. The British in Canada egged the tribes on in enmity to the Americans.

Perhaps worse, the settlers' livelihoods were threatened by the Spanish policy that closed the Mississippi River to American traffic. Only Spanish watercraft could carry the settlers' produce to New Orleans, and the Spanish of course charged a hefty duty on such shipments. The alternative for the settlers — arduous overland routes to the eastern seaboard — was even more expensive, and the settlers saw their existence threatened. There was plenty of seditious talk in Kentucky about becoming an independent republic or a province of Spain, which had what seemed to be a stable government and maybe the ability to establish order in this large and relatively chaotic region. The lands lying west of the Appalachians and east of the Mississippi were plied by spies and hustlers.

Into these murky geopolitics in what amounted to untamed country, George Morgan emerged with his grand plan. He had been a fairly successful Indian trader in the Illinois country before the American Revolution, where he had become a great admirer of the Delaware Indians he dealt with. They, in turn, admired him greatly, chiefly because they saw him as a man of honesty and fairness. As did many white men in this frontier region (and many back east like George Washington and Thomas Jefferson), Morgan also engaged in major land speculation schemes, failing in both of his two attempts. With the outbreak of the Revolutionary War, he became an Indian agent in the Indiana Territory with the rank of captain in the army. (The Indian Department was part of the military then.) Morgan was an outspoken advocate of a policy of neutrality for the Indians and worked hard to achieve it, but the British and his superiors in the American army thought otherwise. Most of the Indians in the Ohio Valley lined up with the British, even — finally — all but one small group of Morgan's beloved Delawares. Disappointed and seeing himself no longer of much use as an Indian agent, Morgan resigned, and, after a stint as a supply officer (with the rank of colonel) at Fort Pitt, he left the army altogether to take up farming in Princeton, New Jersey.[2]

Something of a Jefferson wannabe, he undertook various experimental initiatives, hoping to put American agriculture on a scientific basis, and he lived a pleasant, even rewarding life there with his family for nearly a decade. But the old colony-building bug bit again, and soon he was a leading figure in a New Jersey company formed with an eye to obtaining lands in Indiana. The company's application met with delay and, Morgan suspected, bureaucratic sabotage by the infant government's Board of Treasury. Morgan meanwhile heard that the Spanish minister, Gardoqui, had instructions to promote the development of Louisiana by awarding land grants to those who would become colonists.

Morgan quickly submitted an elaborate plan, many details of which he most likely lifted from his earlier attempts to start colonies in the Ohio Valley. It called for a large section of land extending west of the Mississippi and running north and south of the mouth of the Ohio River. Morgan was to be given command of the colony, which would be self-governing while subject to the king of Spain, and it would enjoy freedom of religion even though all other Spanish colonies were Catholic only. Morgan would see to the building of a port city to rival St. Louis farther north, and one that would command the Ohio and lower Mississippi river trade, providing duty payments to the Crown. Morgan proposed that he be paid for his efforts as well as receive land grants of a thousand acres for himself, his wife, and his five children. He also proposed to name the colony New Madrid in honor of its Spanish sponsors.

Gardoqui was delighted with the plan and with Morgan's evident ability to organize such an undertaking, not to mention Morgan's standing in American society. This was no fly-by-night adventurer, but a man of stature. Morgan's plan for the port city, also to be called New Madrid, was elaborated in great detail. It called for wide streets in a rectangular pattern, sidewalks well above the streets, each block containing six acres "which shall be sub-divided by meridian lines, into twelve lots of half an acre each." Each block would have land set aside for a church, and other land within the city would be set aside for schools and other public uses, including parks. A lake in the middle of the city would be tree-lined, as would many of the city's boulevards, and no one could harm any tree in the city without express permission from the chief magistrate — in other words, Morgan himself. Indeed, the gentleman farmer insisted, the trees and shrubs "shall be religiously preserved as sacred." Harking back to his days with the Delaware Indians, Morgan also insisted that no white man earn a living in the colony by hunting. Colonists could hunt for their own consumption but not as a business. Instead, Indians would be permitted to hunt widely and freely in the colony, for Morgan considered hunting and the trade in pelts to be their chief livelihood.

The Spanish minister was so pleased with this utopian vision that he gave it provisional approval and agreed to help finance

Morgan on an inspection trip while he — Gardoqui — sought final approval from the Louisiana governor, Don Esteban Miró, and the king. Assuming the minister's enthusiasm spelled success, Morgan produced a handbill advertising the colony. Each colonist would be given 320 acres of good farmland for an eighth of a dollar an acre. The colonists' farm products would enjoy free navigation of the Mississippi and a duty-free market in New Orleans. The prospectus was compelling; in January 1789, Morgan and some seventy companions headed down the Ohio from Fort Pitt. By April he had fastened on the site where he would erect his grand city of the central Mississippi. It was a place known as L'Anse à la Graisse, or the Greasy Bend, located on the bank of what we call an oxbow in the river — a narrow-waisted loop where the river, having turned eastward, proceeded *north,* then curved west and south again. The odd name evidently arose from the greasy detritus of the peltry trade run there by two Frenchmen, the brothers Le Sieur.

Morgan and his fellow colonists described the place in glowing terms, overlooking the existence of swamps and marshes that were common along the river, and also disregarding the effects of flowing water on the banks of a river. Leaving the colonists in place, an elated Colonel Morgan traveled downriver to New Orleans to discuss the colony's progress with Governor Miró.

The governor was not at all pleased. One of the spies plying the region was American general James Wilkinson, who was commander of the West and also secretly a Spanish agent. He had obtained from Esteban Miró what he took to be an exclusive right to establish American colonies on the Mississippi's west bank. Before Morgan reached New Orleans, Wilkinson had already denounced him as a dangerous interloper, and the governor was prepared to scuttle Morgan's scheme. Gardoqui, he announced, had far exceeded his authority in making a conditional grant to Morgan. The settlers should get land free, as grants, not pay for it. No colonist could sell land for his own gain, and that included Morgan. He would recommend that Morgan be given no more than a thousand acres in all for himself and his sons. The Spanish military would be in charge of the colony, and a fort would be built in the middle of town. Religious toleration was, of course, out of the question in a colony of His Catholic Majesty, as was any form of democracy. Governor Miró even found Morgan to be presumptuous in naming the colony after Madrid.

Morgan was unable to settle for such terms. Within a year he went back east and the project died. Most of the American colonists who had accompanied Morgan left as well, the colonel returning to the life of the gentleman farmer (on land in Pennsylvania this time). His greatest dream, his grandest plan, was dashed. People drifted in, however, at Morgan's original site and elsewhere along the river. By 1791, New Madrid was just another unprepossessing frontier town with a population of some two hundred people. From the start, however, the river ate away at the bank on which New Madrid was located. A traveler in 1796, Victor Collot, noted that "every annual revolution carries off from one to two hundred yards of this bank; so that the fort, built five years since," was about to be decommissioned before it was destroyed. By 1797, some two hundred houses were scattered here and there. The fort had been rebuilt yet farther inland.

In 1800 the king of Spain ceded the Louisiana Territory to France, and three years later a beleaguered Napoleon sold it to the United States for $15 million, a pittance. New Madrid was now part of the United States of America. By the fateful year of 1811, most of the original town site had disappeared into the river, along with the remains of the fort. Some five hundred people eked out a living in the town and nearby. It was a swampy place of little hope, just another dreary little frontier town. About all that remained of George Morgan's dream city was the name.

The Entrails of the Times

Simply said, 1811 was a very bad year. The spring floods on the Ohio and Mississippi were so high it came to be known as the "year of waters." At year's end a particularly violent hurricane ravaged the eastern coast of North America from Cape Hatteras to Labrador, and a fire in a Richmond, Virginia, theater incinerated some seventy people, including the state's governor. Presiding over these calamities was a comet,[3] and people had long perceived comets as harbingers of bad news, portents of earthly catastrophes. After all, as one can see today in the Bayeux tapestry, a comet (later determined to be Halley's) presided over William the Conqueror's defeat of the hapless Harold at the Battle of Hastings in 1066, decisively changing British history. A fifteenth-century pope called one comet an agent of the Devil. As late as 1910, people still feared that lethal emissions from Halley's comet would poison the earth's atmosphere, and this cosmic interloper was accused ex post facto of causing a terrible flood in Italy. Even Comet Kohoutek, the comet of 1973 that was highly touted as a real dazzler but turned out to be nearly invisible to the naked eye, called forth dire predictions. Back in 1811, many Americans (and others around the world) feared that the comet lighting up their sky might well collide with the earth with the most appalling consequences. Then as now, doomsayers and other superstitious folk were sure the comet presaged the end of the world.

This was not the case, of course, in the world's small community of astronomers, an American member of which has been credited by historians with discovering the comet of 1811. William C. Bond was at the time a young scientist noted for developing a ship's chronometer. After discovering the comet in 1811 he went on to become the first director of the Harvard Astronomical Observatory, discovering the innermost ring of Saturn and its

eighth satellite, Hyperion, as well as making the first clear da-
guerreotype of the moon. A distinguished career, to be sure, but
Bond was not the discoverer of the 1811 comet. Transatlantic
communication being relatively slow, he may have been unaware
that a French astronomer named Honoré Flaugergues had already spotted the comet, which now, in the scientific annals if not
the *Encyclopaedia Britannica,* bears the name Flaugergues.

Outside the astronomy community, many fretted about dire
effects arising from the comet and a host of other natural events
that year, including a rash of earthquakes reported from around
the world, and on September 17, people throughout the trans-
Appalachian United States were treated to an eclipse of the sun,
which brought forth from some quarters yet further prophecies
of dreadful events in the offing.

At this time, Americans were no strangers to earthquakes —
more than 150 quakes had been felt in the eastern half of North
America from the time of the Puritans' arrival in the early seventeenth century. One of the worst of these had occurred within living memory of those who experienced the New Madrid quakes:
this one had hammered Boston on November 18, 1755, and was
felt over an area of three hundred thousand square miles from
Nova Scotia to the Chesapeake Bay. A ship some two hundred
miles offshore experienced a huge shock as if it had run aground.
One peculiarity of this quake was that it appears to have killed
vast numbers of fish along the seacoast.

Only eighteen days after the Boston quake, one of the most
devastating quakes in history laid Lisbon, Portugal, to waste and
haunted the minds of people on either side of the Atlantic for
years to come. The Lisbon quake consisted of three rapid-fire
shocks, one of which weighed in at Magnitude 9. The entire Ibe-
rian Peninsula was severely damaged, the mayhem extending
across the Strait of Gibraltar to North Africa. Tremors were felt
1,850 miles away in Sweden. In all, an estimated sixty thousand

people were killed. The Lisbon quake's three huge shocks all occurred within about two hours on November 1, All Saints' Day, and fewer would have died if so many of the faithful had not been in church that morning: all such buildings in Lisbon were destroyed. The first shock (the one that has been shown to have been a 9) lasted for some seven minutes. Among other terrifying effects, including fires that destroyed what the shock had not, the sea itself retreated, emptying the harbor like a bathroom sink, and then roared back to engulf what was left of the city in fifty-foot waves. In addition a huge ocean wave, called a tsunami, swept out into the North Atlantic, causing twelve-foot waves 3,540 miles across the ocean in Antigua.

That the Lisbon quake had struck on All Saints' Day, a solemn festival of the church, was not lost on people. The devout took this calamity as clearly the wrath of God enraged by the imperfect virtue of the people — indeed, like Sodom and Gomorrah, Lisbon had to be destroyed. The Inquisition looked all the harder at its less obedient sheep. Nor was it only the God of the Roman church who had expressed outrage: John Wesley, founder of Methodism, wrote at the time that there "is no divine visitation which is likely to have so general an influence on sinners as an earthquake."

A more secular philosophical debate of the day, represented by two French philosophers, Jean-Jacques Rousseau and Voltaire, also received a jolt from the Lisbon calamity. Rousseau, like the English poet Wordsworth, was of the "divine nature" school of Enlightenment thought, seeing a life lived simply and close to nature as the most virtuous: this was humanity and nature at peace with each other. Nature was, in other words, good. Voltaire, of course, thought otherwise, and in 1756 he wrote an attack on the romantic Rousseau's notion of good and evil. It was called *Poem upon the Lisbon Disaster.* He insisted that there was physical evil in the world, that nature was not necessarily at peace with humanity, nor could mankind expect any protection from divine providence. "We launch ourselves," he wrote, "like missiles, at the unknown."

The rise of Western science has since dismissed earthquakes and other natural events as the acts of angry deities, and has persuaded many that nature, as such, is without morals but instead simply occurs (while evil is an exclusively human province). But many people still feel the presence of Evil as an active, even palpable force loose in the world. The Rousseau-Voltaire debate has never really evaporated. It has undulated through the writings of philosophers like Hegel and Marx (on Rousseau's more optimistic side, suggesting that human brutality can be engineered out of existence), and Hume and others who call for people to resign themselves to a world full of inexplicable evils. The debate over such large questions fell out of favor in academic philosophy departments for many years of the twentieth century, with philosophy becoming more like a technical manual of linguistic constructs. But the debate continues to arise in the pronouncements of certain national leaders — and philosophers are returning to it. Philosopher Susan Nieman reviews the long history of this debate and the quest for reason in the world in her 2002 book *Evil in Modern Thought,* which begins, in fact, with the Lisbon earthquake.

In the aftermath of the New Madrid quakes, some people dredged their memories for other portents, other events that, in hindsight, might have forecast the terrible quakes. People had felt tremors in Columbia, South Carolina, and Augusta, Georgia, as early in the year as January. Some people remarked on the huge roosts of passenger pigeons and Carolina parakeets that had descended on the Ohio Valley (though such huge congregations of these soon-to-be-extinguished birds were by no means rare). In 1835, one person recalled a vast lemming-like migration of squirrels — millions of squirrels "obeying some great and universal impulse" — moving west through Indiana, many perishing in the Ohio River, an altogether improbable memory. Less unlikely (though unconfirmable) is the lore passed down by Allen Holloman of New Madrid, who was five years old when the quakes occurred. To this day, his descendants speak of nervous dogs and cats, horses running wildly, and snakes ignoring the winter cold to rise from their holes and lie on the ground. Wolves and deer, others reported, lost their fear of humans.

A Cincinnati physician, Daniel Drake, was puzzled by the odd appearance of the sun's light in the region in the spring of 1811. Others reported a curious lack of thunder in the region during the year. Drake also recalled an especially severe outbreak of fever among the people who lived along the Ohio and Mississippi rivers. William Leigh Pierce, on his way from Pittsburgh to New Orleans, wrote that on the last day of November 1811, just before sunrise, "two vast electrical columns shot up from the eastern horizon, until their heads reached the zenith." They lit up the sky and came and went in an instant, and thereafter until December 16, the air was never transparent but instead quite opaque. When the sun was visible, "it exhibited a dull and fiery redness."

No actual predictions of the New Madrid quakes, however, have been noted — except those made by one of the most famous American Indians, the legendary Shawnee named Tecumseh. His name in the highly symbolic Shawnee language meant "I cross the way" but also "Panther about to spring." He belonged to the panther clan, whose spiritual progenitor was a sky-borne panther. Yet another meaning of Tecumseh was "shooting star." In 1811, when he was at the peak of his persuasive powers and most successful at rallying Indians from numerous tribes to oppose the loss of their lands to the white man, Tecumseh became associated with the comet that, like a brilliant panther, was crossing the sky, heralding trouble in the very bad year of 1811.

Tecumseh was a well-traveled man. His people, the Shawnees,

lived in the Ohio Valley, in the south among the Creeks and even into Florida, and across the Mississippi in today's Arkansas as well. When Tecumseh took up the goal of uniting the tribes west of the Appalachians into a grand force that would, if necessary, drive the white man out of these lands so that a pan-Indian nation could be established, he traveled far and wide seeking support. Cobbling together such an alliance of Indian tribes over a large geographical area was made all the more difficult by the fact that most tribes simply did not get along, indeed were often engaged in intertribal hostilities. But Tecumseh was spectacularly persuasive as an orator, a man of great dignity and spiritual charisma whom even his enemies would lionize after his death as something of a military genius as well.

This was a time of religious revivalism among both whites and Indians, and the Indian world was full of nativist prophets and seers, most of them perceiving a world soon to come in which the white man would have disappeared and the Indian people could go back to their old pure ways. This too was part of Tecumseh's charisma, for he was considered a seer as well, but his appeals were not always met with unanimous agreement that the gods were angry at the Indians for not repelling the white man.

At an ancient Creek village in Alabama, called Tuckhabatchee, Tecumseh gave his war talk and handed out wampum, a hatchet, and other artifacts symbolic of war, but noticed that the chief, known as Big Warrior, was reluctant to go along with Tecumseh's call for an uprising. He is supposed to have pointed at the old chief and said, "You do not believe the Great Spirit has sent me. You shall know. I leave Tuckhabatchee directly, and will go straight to Detroit. When I arrive there, I will stamp on the ground with my foot and shake down every house in Tuckhabatchee."

In fact, Tecumseh did not go straight to Detroit but stopped off at various places on his way north. But Big Warrior and others kept his threat in mind. On the day they had fixed as the likely time of Tecumseh's arrival at Detroit, they awoke to a great rumbling of the earth. The ground shook like a blanket and, indeed, all the houses in the village collapsed. It was December 16.

So much of Tecumseh's life is enveloped in legends that it is difficult to separate fact from fancy in many details. Whether Tecumseh actually made this prediction or not, however, the fact is that many Indians believed he had predicted the terrible quake and that it did indeed represent the anger of the Great Spirit over the depredations of the white man and the failure of the Indians to rise up. Adherents flocked to Tecumseh's cause. Later, after the New Madrid quakes, Tecumseh used them and his "prediction" to rally yet more forces. While he was among the Osages in Missouri, a white captive named John Dunn Hunter recalled him fulminating that the Great Spirit was "angry with our enemies; he speaks in thunder, and the earth swallows up villages, and drinks up the Mississippi. The great waters will cover their lowlands; their corn cannot grow; and the Great Spirit will sweep those who escape to the hills from the earth with his terrible breath."

At one point in his career, Tecumseh managed to bring three thousand Indian warriors into a battle, the greatest aggregation of warriors in all of Indian history. But his crusade was, it turned out, too late. He played a crucial role in the British defeat of Detroit once the War of 1812 broke out, but was killed in a battle thereafter in Canada. He was irreplaceable; no other Indian was capable of holding all the different tribes together, and his dream of tribal unity was shattered. Even though the Great Spirit had spoken — thunderously — through the violence of the New Madrid quakes, the Indian would be on the run for the rest of the century.

A Question of Honor – and Land Hunger

Perhaps even more ominous for people in the United States in 1811 were the presentiments of war that hung heavily over the United States. For the previous six or so years, while the great European powers had been engaged in complex wars and kaleidoscopic alliances, British and French policies were designed to interfere with American trade, even going so far as to sanction the seizure of American ships and American seamen. By the onset of the winter months, Americans, particularly those in the southern and western states and territories including the area that the great quakes had most affected, were baying for open war that would reclaim American honor, pay back the British for fomenting and supplying the Indian attacks on the western frontier, and also fulfill a desire to grab Florida (which at the time included the southern parts of today's Alabama, Mississippi, and Louisiana) from the Spanish. The great prize would come from snatching Canada from the British.

On Capitol Hill, bombast flourished. John C. Calhoun, an ardent nationalist who would later become one of the most outspoken Americans to ever take up the cause of states' rights, fulminated for war, claiming that the nation was "roused from its lethargy . . . determined to vindicate its honor." He was most volubly opposed by John Randolph of Roanoke, who declaimed that if America took on such an adventure it could not escape the calamities "providence has thought fit to inflict on those nations which have ventured to intermingle in the conflicts now going on in Europe." Later, in language that suggested he was a believer in the comet as a harbinger of troubles (but which was probably just standard congressional hyperbole), Randolph said entering into a war with Britain would heap on the nation "a dreadful scourge, some great desolation, some awful visitation." With ample justification, historian James Penick, Jr., said of these exchanges that, like "the immoveable object and the irresistible force, inflated eloquence met magnificent pomposity."

Finally, in June of 1812, even as the comet disappeared from sight (for another 3,094 years, we now know) and tremors continued to shake the ground and rattle nerves virtually every day in the region surrounding New Madrid, the United States declared war on Great Britain. Soon the greatest maritime power on earth, master of some six hundred warships, was faced by a shaky new nation with a fleet numbering sixteen. The United States government had far more pressing matters to worry about than what it had learned about the New Madrid quakes . . . with one exception.

On March 26, 1812, less than two months after the hard shock at New Madrid, a huge and particularly vicious earthquake struck Caracas and La Guaira in Venezuela. Both cities were for the most part destroyed. It was Holy Tuesday, and many of the twenty thousand people who died in the quake were buried in churches. The Catholic priests of the region considered this to be a sign of the wrath of God and sought to call off the revolution under way at the time under the leadership of Simón Bolívar, the great liberator of South America. This caused a temporary setback for the revolutionaries, and a sympathetic United States Congress sent $50,000 to help with the recovery of the stricken cities.

Soon enough the territorial assembly of Missouri complained, suggesting that Congress show similar compassion to the people who had suffered in the New Madrid quakes. Thus, in 1815, Congress passed a disaster relief act — the first such act in the nation's history — and almost immediately Congress had a lesson in the unintended consequences that can arise from even the most well-meaning legislation.

The act stipulated that people whose land had been devastated by the New Madrid quakes could trade it for unclaimed land elsewhere in the Missouri Territory, up to 640 acres (one square mile, and what was called in homesteading days a "section"). One consequence of this was that real estate speculators and dealers from St. Louis and the East heard about all this before the people of the New Madrid area. Many of the locals were soon conned into selling their "worthless" land to the speculators (in some cases for as little as sixty dollars for an entire section), who in turn traded it in for far more valuable land elsewhere. The official instruments for these arrangements were called New Madrid Certificates, and before long some of the rural people got wise, duplicating their certificates and selling off their places to as many as ten hapless con artists. The resulting tangle would tie up Congress and the courts for decades and never did get satisfactorily resolved. New Madrid Certificates became as common a synonym for fraud in their time as the appeals via the Internet from Nigerians needing a bit of money laundering are today.

3

Pendulums and Polymaths

AT THE TIME of the New Madrid quakes, there was little by way of geological knowledge or theory by which anyone could have understood what had caused them. Most people who thought of themselves as scientists still believed generally in the history of the world as specified in the book of Genesis. The Old Testament's begats had been stretched backward in time some six thousand years by an Anglican, James Ussher, who was the archbishop of Ireland. Creation, according to Ussher, had occurred on October 23, 4004 B.C. (a Sunday). Another prelate soon pinpointed the minute as nine o'clock in the morning. Not long after the Creation, an array of catastrophes — especially the Flood — had brought the earth as known in the nineteenth century into being. It was the Flood, for example, that had managed to roll huge boulders from the north to the heights of the Alps (the Ice Ages were yet to be discovered). Two main schools of thought in the inchoate realm that would become geology existed on the question of what lay beneath, or before, the catastrophes in the first place: Neptunists and Vulcanists. (Neither group was wedded to Ussher's precise date.) The Neptunists, also known as Wernerians for this school's founder, Abraham Gottlob Werner of Freiburg, believed in an aqueous origin of rocks by which bits of material had precipitated from a primeval ocean and, in deposit after deposit, had built the structure of the earth. In a sense, they believed the world to be formed of sandstone and shale and similar rocks. The other school was known as Vulcanists, and also called Huttonians after the Scot, James Hutton, who was among the first to suggest that the same processes going on in the present, such as erosion, had shaped the earth's crust over a time scale far greater than the Bible permitted. This was called Uniformitarianism. Vulcanists believed that the structure of the earth came about as a result of subterranean heat and was formed mostly by volcanic rocks like basalt.

The great Scottish geologist Charles Lyell (who would eventually visit New Madrid) was yet to begin publishing his three volumes that formed the very platform of modern geology, establishing Uniformitarianism à la Hutton as the essence of geological change rather than the Catastrophism preferred by those wedded to the biblical view. Through his work, the first volume of which appeared in 1830, the time frame for geological history was extended far enough backward to make room for Darwin's insights into the evolution of life forms, which he published in 1859. As Darwin wrote of Lyell's work, "it altered the whole tone of one's mind, and therefore . . . when seeing a thing never seen by Lyell, one yet saw it partially through his eyes."[1] Of course, many people with a devout regard for the Bible were scandalized: so revolutionary was Lyell's new time scale that for several years women and children were excluded from his lectures.

But at the time of the New Madrid quakes, Lyell was far from publishing his ideas, and the overall state of geology in the new

United States was indeed some ten or even twenty years behind that practiced in Europe. The first colored geological map of the United States (meaning the lands from the Mississippi to the Atlantic) was published in 1809, drawn up by a wealthy immigrant Scot, Willam Maclure, who took up geology in midlife and saw his map, though crude by even contemporary standards, as providing a key to the new nation's economic development. The map was accompanied by Maclure's *Observations on the Geology of the United States,* published in the *Transactions of the American Philosophical Society* and based on his many treks through the Appalachian countryside.

Something of a visionary, Maclure later hooked up with fellow Scot Robert Owen, the famous social reformer, and joined his utopian community, New Harmony, in Indiana. In 1826, with a flair for the dramatic, Maclure brought to New Harmony a group of prominent intellectuals aboard a keelboat christened the *Boatload of Knowledge.* When the experiment in community living failed, Owen returned to Scotland and a disappointed Maclure went to Mexico, where he died a few years later.

Maclure was a Wernerian, and his map simply excluded such rock forms as basalt and other igneous and volcanic rock. When he revised the map in 1817, he conceded that basalt and granite were volcanic, and the "Neptunian" rocks were reduced but not altogether eliminated. Still, Maclure's map added to what was a very slim, almost nonexistent, database, and in honor of his pioneering effort to create such a map, Maclure has been referred to as "the father of American geology."

With the science of geology in its infancy, it is little wonder that the ideas available to explain earthquakes and in particular the New Madrid quakes appear today to be so quaint. One notion, already mentioned, was an imbalance in the earth's electrical fluids. Some connected the New Madrid quakes with the quake that devastated Caracas in Venezuela in 1812, speculating that a vast fault stretched from mid-America across the Caribbean and into South America. Thomas Nuttall, a British naturalist who accompanied John Bradbury on part of his trip up the Missouri River (and unlike Bradbury, remained in America for the rest of his life), wrote in 1817 that a "vast bed of lignite or wood-coal situated near the level of the river, and filled with pyrites, has been the active agent in producing the earthquakes" at New Madrid. He said that the time will surely come when the world will witness "something like volcanic eruptions on the banks of the Mississippi."

Another geologist, Edward M. Shepard of the United States Geological Survey, wrote that "it is difficult to make any statement. It may have been due to the readjustment of the fault lines in the Ozarks, or to a similar cause in the Appalachians." Shepard went on in a Neptunian vein to say that "there can be no doubt that it was due to great artesian pressure from below," removing material below so that a "slight earthquake wave would destroy the equilibrium of a region thus undermined, resulting in the sinking of some areas and the elevation of others." In the absence of a great deal of information yet to be developed about the nature of the earth, all suggested causes of earthquakes were at this time sheer speculation.[2]

Then as now much of science was a matter of measuring, putting numbers on things, categorizing phenomena, and the measuring instruments available for such work were pretty crude. As far as earthquakes are concerned, what was known about them in the early nineteenth century was almost entirely anecdotal — in the form of eyewitness accounts. There were very few of those items so necessary to doing real science and so dear to every scientist's heart — numbers. But at least one person made a systematic attempt to count and measure the New Madrid quakes.

Practically nothing is known about him, in spite of the fact that he may well be the first truly scientific seismologist in the United States.

The Mysterious Jared Brooks

Jared Brooks was considered a talented if eccentric engineer and a surveyor. He lived in Louisville, Kentucky, and died in 1816, four years after the last shock of the New Madrid quakes.[3] In December 1811, Brooks began to keep track of the shocks that had originated to the south in the New Madrid area, and he continued until well into March of the following year. Brooks described the first shock as felt in Louisville: "A great noise is produced by all the loose matter in town, but no sound [of the earthquake itself] is heard; the general consternation is great, and the damage done considerable; gable ends, parapets, and chimneys of many houses are thrown down."

Of greater interest, however, was his description of what he perceived to be the mechanisms underlying the shock and its effects. "It seems," he wrote, "as if the surface of the earth was afloat and set in motion by a slight application of immense power, but when this regularity is broken by a sudden cross shove, all order is destroyed, and a boiling action is produced, during the continuance of which the degree of violence is greatest and the scene most dreadful; houses and other objects oscillate largely, irregularly, and in different directions." Clearly, Brooks had some means of making note of the different motions perpetrated by this earthly seizure that was taking place some 230 miles away. He had adorned some part of his rooms with pendulums of various lengths — from one to six inches — and a weight suspended on a spring. With these he could monitor ground motions, even those not perceptible to the human senses, the pendulums showing horizontal motions and the spring vertical motion.

We are all familiar with the conventional pendulum that swings back and forth in, say, a clock that itself sits solid and motionless on a mantelpiece. But imagine the clock sitting on a table with its pendulum at rest. If you were suddenly to jerk the clock to the side, the inertia of the pendulum bob would cause it to remain stationary for a moment and then swing. Its swinging lags behind the motion of the clock itself, and this relative motion can be recorded in the form of a wiggly line by a pen on paper. Similarly, if you attach a weight of some kind to a spring that is attached to a frame anchored to the ground, any vertical motion of the ground will make the frame move up and down relative to the suspended weight, and this motion too can be recorded.

Such is the basis of today's more sophisticated seismographs, which now record the motions electronically. A seismograph, technically speaking, not only detects the earth's motions but also records them. Such a record — often in the form of a jagged up-and-down line on graph paper though today a computer printout — is called a seismogram.

A device that only detects such motions without recording them is a seismoscope. If you have only a seismoscope, it is necessary to sit patiently for hours on end watching the instrument to see when an earth tremor occurs. Brooks's pendulums and spring amounted to seismoscopes, and in recording the shocks and aftershocks of the New Madrid quakes from December through March of the next year, he must have spent hour after lonely hour staring at his primitive instruments.

Seismoscopes of various kinds of a far less useful type than those of Jared Brooks go back a long way, however. The first one known to history was developed in A.D. 132 by the Chinese philosopher Zhang Heng, and it was a far grander instrument, at least aesthetically, than Brooks's pendulums. It consisted of a closed bronze urn — six feet in diameter — with eight bronze dragons each holding a bronze ball in its mouth, ranged around the urn's circumference. A bronze frog sat beneath each dragon. When a sufficiently strong tremor of the earth occurred, a pendulum inside swung and opened up the jaws of the dragon facing the oncoming tremor. This in turn let the ball fall into the mouth of one of the frogs. This recorded the direction of the tremor by virtue of which frog had a ball in its mouth, but it left no record of the severity of the tremor.

Zhang's elaborate machine was evidently more of a curiosity or toy than an instrument for collecting information, and it must have had some kind of free-swinging pendulum inside to achieve its results. Millennia later, a Frenchman, J. de la Haute Feuille, filled a bowl with mercury and arranged eight smaller containers around it on the assumption, widely held at the time, that the earth tilted during earthquakes. The idea was that the mercury would spill out of the large bowl in the direction away from the source of the movement. But since most earthquakes cause horizontal movement, not much came of Feuille's mercury. In the early eighteenth century, Nicholas Cirillo of Naples used a pendulum to record the amplitude (or length) of pendulum oscillations at two sites when some severe shaking of the Naples area took place.

Jared Brooks was no doubt aware of the developments in Italy, as evidently were others in Louisville at the time. A chronicler of the times said that many Louisville residents would hang a pendulum in one of their rooms and, if it appeared to be indicating a severe shock, they fled.

Brooks's most heralded results were the number of shocks he counted — from December 16 until March 15, a total of 1,874. This was an amazing number. That quantity of shocks was unheard of. For example, the 1773 earthquakes in Calabria — the southern "toe" of Italy's "boot" — consisted of six huge shocks and a large number of smaller ones, or aftershocks, that devastated the area, leaving few houses standing and killing thousands of people. In the four years afterward, a physician known only as D. Pignataro of Monteleone felt 1,186 aftershocks, about the same number Brooks recorded in three *months*. Of course, Brooks was counting shocks that were too slight to be felt by human senses as well, but, even so, what was happening along the Mississippi River was something unprecedented. Today's far more sensitive seismometers would surely have picked up more tremors than Brooks's crude instruments. It is likely that the actual number of tremors was at least twice what Brooks recorded, meaning some 3,600, or, on average, 40 a day for the three months.

The differing lengths of Brooks's pendulums permitted him to perceive the differences in amplitude of the vibrations brought on by various shocks, as well as their rapidity. When the amplitude was short, only the one- or two-inch pendulums vibrated. At other times the three- or four-inch ones moved while the longer ones and shorter ones remained still. Brooks kept a thorough log of these events as well as the weather, which he may or may not have associated with them. For example, on February 1, a relatively quiet day, he wrote, "Morning cold, hard frost; light and broken clouds; wind south-east; sun shines dimly at times. 9 h. A.M. — Tremors and slight vibrations commence and continue with but short intervals until about 1 hr. P.M.; at that time fair sun through an atmosphere of whitish complexion; evening somewhat hazy; not a breath of wind."

On February 5, after noting that it was one of the fairest mornings of the winter with one and a half inches of ice produced during the night, he described a shock of about one minute at eight-thirty in the morning, with tremors following. At quarter past one in the afternoon, he wrote, "Vibrators act perpendicularly and strongly nearly one minute, whilst the [pendulums] scarcely move, then all get into strong action; continuance of this shock (as it may be termed) two minutes." That night it was "severely cold, fair, and dead calm." A thin vaporous overcast took the sky at about eleven o'clock when the assiduous Brooks apparently gave it up for the day.

Two days later on February 7, the morning of the hard shock, after noting the date, he wrote, "3 h. 15 m. A.M. the most tremendous earthquake yet experienced at this place, preceded by frequent slight motions for several minutes, duration of great violence at least four minutes, then gradually moderated by exertions of lessening strength, but continued a constant motion more than two hours, then followed a succession of distinct tremors or jarrings at short intervals, until 10 h. A.M. when, for a few seconds, a shock of some degree of severity, after which frequent jarrings" and so forth through the rest of the day and the evening until about midnight when he noted that it was cloudy, calm, and some snow on the ground was melting. Keeping these records could be tedious work, one has to guess from a comment Brooks made about a severe shock and its aftermath the evening of February 7 after ten o'clock: "The last shock was violent in the first degree, but of too short duration to do much injury, subsided suddenly, and is followed by constant trembling . . . at intervals till one is tired of counting."

It is not at all clear what, if anything, Brooks planned to do with the data he collected, but they were not published before his death. They might very likely have been lost to history and to science had it not been for a medical doctor from Philadelphia who arrived in Louisville in 1816 and promptly set out to write the history of the town. Henry McMurtrie may earlier have been a prisoner of the British for two years, the ship on which he served as surgeon having been captured. With the war over, he looked westward, alighted in Louisville, and began at once collecting information for his *Sketches of Louisville*. He obtained the records of the New Madrid quakes from Brooks, whom he evidently ad-

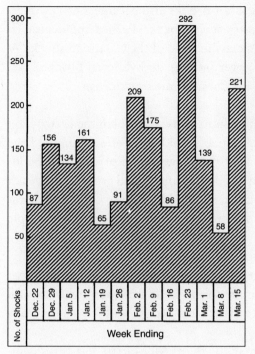

Figure 4. Earthquake activity by week, based on Jared Brooks's summary of the shocks at Louisville, Kentucky. Derived from Myron L. Fuller, "The New Madrid Earthquake," *U.S. Geological Survey Bulletin*, 494, 1912.

mired, and included them in the appendix of his book, which was published in 1819. Then, in debt to both the printer and his landlord, McMurtrie left town, returning to Philadelphia, where he practiced medicine, translated the zoological work of the great French scientist George Cuvier, and was later hired by John James Audubon to edit and oversee the publication of his landmark, multivolume work on the birds of America. He went on from there to be a teacher and an active member of the Philadelphia Academy of Natural Sciences.

Withal, McMurtrie's lasting contribution is the rescue of Jared Brooks's earthquake data, which represented more than merely a careful record of the shocks and aftershocks of that titanic event, as useful as that has been. For Brooks also became the first American to attempt to classify such shocks. (Earlier, the doctor D. Pignataro had essayed a classification of the Calabrian shocks — slight, moderate, strong, very strong, and violent.) Brooks's system was more specific. He classified the tremors as:

First rate. Most tremendous, so as to threaten the destruction of the town, and which would soon effect it should the action continue with the same degree of violence; buildings oscillate largely and irregularly and grind against each other; the walls split and begin to yield; chimneys, parapets, and gable ends break in various directions and topple to the ground.
Second rate. Less violent but very severe.
Third rate. Moderate but alarming to people generally.
Fourth rate. Perceptible to the feeling of those who are still and not subject to other motion or sort of jarring that may resemble this.
[Fifth rate. Not defined. Perhaps McMurtrie lost this definition or merely forgot to include it.]
Sixth rate. Although often causing a strange sort of sensation, absence, and sometimes giddiness, the motion is not to be ascertained positively, but by the vibrators or other objects placed for that purpose.

Brooks was not the only person of scientific leanings who took note of the New Madrid quakes. In 1823, another of the multitalented folk who turned up here and there on the frontier, John Haywood, published *The Natural and Aboriginal History of Tennessee*. Haywood was what we would think of as an amateur archaeologist and historian but was at the same time a lawyer who served as a justice of the supreme courts of both

North Carolina and Tennessee. He wrote of the New Madrid quakes:

> The motions in Tennessee were sometimes, but seldom, perpendicular; resembling a house raised, and suddenly let fall to the ground. Explosions like the discharge of a cannon at a few miles' distance were heard; and at night, flashes of lightning seemed sometimes to break from the earth. For two or three months the shocks were frequent; almost every day. Then they gradually decreased in frequency and took place at longer intervals, which continued to lengthen till they finally ceased. In May 1817, in Tennessee, they had come to be several months apart and were but just perceptible. The last of them was in 1822.

Without Jared Brooks's pendulums and patience, most of the shocks after March 1812 presumably went unnoticed, being beyond the human sensorium. Today we know that the shocks have gone on, usually beneath the radar of most human senses, until today. Brooks is credited with being the first American to employ such instruments for so serious an investigation of earth tremors, thus making him in some eyes at least the father of American seismology. Science historians like to assign paternity in the inchoate investigations that one day lead to an actual branch on the great tree of science. But choosing such sires can be a tricky business: almost every branch of science really has multiple sires, and so does this one. Certainly, however one assesses such things, Brooks was one of scientific seismology's earliest pioneers.

Daniel Drake, a doctor in Cincinnati, brought scientific acumen equal to that of Jared Brooks to the earthquakes that rattled Cincinnati as they did Louisville. And Dr. Drake evidently used pendulums to detect the shocks too light for all but "the delicate

sensations of a few nice observers." Like most doctors of the
time, Drake's education included chemistry, and also like most
doctors, he was a pharmacist as well or — as they say in England
to this day — chemist. Drake, whose medical education began in
1804, started a program of medical education over a drugstore in
Cincinnati and went on to found the College of Ohio (the nation's tenth oldest medical college), Cincinnati College, the Commercial Hospital and Lunatic Asylum, and the *Western Journal
of the Medical and Physical Sciences*. A man of science and a naturalist, Drake is quite properly memorialized in Daniel Drake
Park in Cincinnati.

In a book he wrote about his adopted city, *Natural and Statis-
tical View, or Picture of Cincinnati and the Miami Country*,
published in 1815, Drake also categorized the quakes — into
five classes, the first being the three major shocks. The second
and third classes were shocks of descending violence. The fourth
class, which comprised half the shocks, were those felt only by
people who were at rest. The fifth were those "tremors & ebul-
litions" detected by pendulums and "nice observers." Drake
noted that the focus of the shocks originally was between New
Madrid and Little Prairie but, as the years went by, they moved
northward and then up the Ohio River. He noted as well that,
unlike the undulations of the earth common on the Mississippi,
the shocks felt in Cincinnati were vertical. "The cause," he
wrote, "acted directly upwards, and elevated to the surface of the
earth, sand and various extraneous fossils, which had been buried in the alluvium of the river for unknown ages."

As did others, Drake drew attention to the fact that the water
geysering up out of the ground in the numerous sandblows was
warm. Another curious fact he reported was that the shock of
December 16, 1811, "seems to have been stronger in the valley
of the Ohio, than in the adjoining uplands. Many families living
on the elevated ridges of Kentucky, not more than 20 miles from

the river, slept during the shock; which cannot be said, perhaps, of any family in town." This particular anomaly of the quakes would be explained about a century and a half later when the nature of seismic waves was more clearly understood.

Experimentally, Drake ruled out electrical causes of the quakes — at least to his satisfaction: "On the 6th of February 1812, I had a pointed iron rod, supporting a cork ball electrometer, inserted six or eight inches into the moist earth. It was faithfully observed during two of the shocks which occurred in the night of that day, but not the slightest electrical appearance was perceptible." On the other hand, Drake (like Brooks and others at the time) believed that weather conditions were somehow associated with the quakes, for he devoted considerable space in his appendix to the state of the atmosphere, saying that "the following conclusions are deducible":

1. The principal shocks were preceded by an increase of atmospheric heat.
2. They were preceded and succeeded by a *south-east* wind.
3. They were attended with a hazy, turbid or cloudy atmosphere.
4. They, as well as many belonging to the second and third classes, occurred when it was calm, or nearly so, and were succeeded by stormy weather . . .
5. The smaller vibrations of the third and fourth classes, happened in various states of wind and weather.

Drake's preoccupation with the weather as somehow connected to earthquakes seems quaint today, even a bit laughable, but it is rarely wise to laugh at the ignorance of earlier generations and their strange and seemingly irrelevant notions. For today will be succeeded by tomorrow, at which point today may seem laughable too. In Drake's time there simply was no accepted explanation for earthquakes, and we still today are not sure what caused the New Madrid quakes.

In 1814, another medical doctor put together an account of the New Madrid quakes "in the expectation that physical occurrences so immediately before our eyes and under our feet, would have qualified me to form something of a tolerable theory of earthquakes." This was Samuel Latham Mitchill,[4] who was quite possibly the most versatile American man of science of his time.

Taught at the university in Edinburgh, he knew virtually everything there was to be known in science of the period. For in addition to medical practice and a professorship in the College of Physicians and Surgeons at Columbia College in New York City, Mitchill was a chemist, a student of contagious disease, an ichthyologist who produced the first study of the fish of New York, an agronomist and botanist, and a mineralogist and geologist who produced the first descriptive geology text of North America (sharing siredom of that field with Maclure). He was author of numerous scientific and medical treatises as well as the first guidebook to New York City, amused himself by speaking in verse, and founded the nation's first medical journal, the *Medical Repository*, in 1797, along with a number of learned societies that exist to this day.

If that were not enough for a career, he also served in the New York State legislature for three years, and spent twelve years beginning in 1801 in the U.S. House of Representatives and the Senate. As his biographer noted, "The Congresses of 1801–1813 were not overstocked with learned men, and their astonishment, and probably dismay at finding a scholar among them, can be easily imagined" (a comment with continuing applicability, it seems). In fact, Mitchill was much given to the pedagogical arts, often turning a speech on the floor into a lecture. Once, for example, when the subject of congressional debate was preparations for war, the Doctor — as he was called — drifted off into a learned talk on the history of making gunpowder and the chemical affinities of bodies. Despite what may have struck a long-winded age as logorrheic, Mitchill was President Thomas Jefferson's "Congressional Dictionary," and to many the Nestor, *the* wise man, of American science.

But the wise man admittedly met his match when confronting the cause of the New Madrid quakes. Soon after the major shocks, Mitchill began collecting accounts, which he assembled into an address to a learned society in Washington, D.C., in the spring of 1814. He took note of many kinds of effects. For example, it was clear that the great shock that occurred on December 16 was felt almost simultaneously throughout the countryside from Washington, D.C., to South Carolina and north into Ohio. He recounted light effects: in Louisville, Kentucky, the night had been very dark but after the first shock, it was light enough to "see a pin." Here and there the shocks were accompanied by a sound that most of its hearers compared to that of carriages rolling over cobblestone streets, but sounds did not accompany the shocks in every instance or place. In some cases the shocks seemed to have come from the southwest and headed for the northeast, in other places the reverse.

He cited a report from Columbia, South Carolina, that the college there had shaken so badly with the first shock of December that the alarmed students left their chambers without their clothes (an event that may well be considered the sire of campus streaking). Far to the north, near Detroit, a small lake with an island inhabited by Indians began to "boil like a great pot over a fire; and immediately a vast number of large tortoises rose to the surface, and swam rapidly to shore, where they were taken for food."

The January 23 shock seemed to Mitchill to be more severe than the December 16 shock, since according to the information he had at hand it was felt as far north as New York. (The earlier one seemed at that time to have been confined to the country south and west of Maryland. It was only later that reports emerged of the tremors in Quebec.) In Richmond, it was supposedly "more sensible on . . . hills, than in lower parts of the city," — the reverse of such findings elsewhere.

Young Pierce, the writer who was journeying on the river during the December quake, had collected some of the coal that was ejected along with the warm water from the earth and sent it to Mitchill. "I found it to be very flammable," Mitchill wrote; "it consumed with a bright and vivid blaze."

After reciting numerous eyewitness accounts, and summarizing the multiple effects reported, Mitchill said to his 1814 audience that whatever hypothesis one might believe in "to explain the awful phenomena of earthquakes," one could find support in the reports of the New Madrid quakes:

> The *mechanical* reasoner will find the great strata of the earth falling in some places, rising in others, and agitated everywhere. The *chemical* expositor will discover evidence enough of subterranean fire in the coal, hot water, vapor, and air bubbles which were ejected and extricated. The *electrical* philosopher will deduce from the lights, the noises, and the velocity of their motions, conclusions favorable to the origins of earthquakes from electron, that subtile and universal element. Even the believer in the conversion of metallic potassium, by rapid inflammation, into common potash in the deep recesses of the earth, will find . . . a better argument than any I am acquainted with, to countenance the *alkaline* system of earthquakes.

Mitchill concluded his address by throwing doubt on any and all such explanations. As "plausible, in some respects, as each of them is, [they are] deficient in that general character and universal application which ought to pervade scientific researches."

Thus was the last word spoken for the next several decades about the state of earthquake science in America.

Mitchill also wrote of his hope that the information he had assembled would one day "assist some happy inquirer into nature, to deduce a full and adequate theory of earthquakes." Almost two centuries later in 2002, Susan E. Hough,[5] then president of the American Seismological Society, took on the widely held view among scientists that anecdotal accounts of such things as earthquakes are suspect to the point of being useless. Speaking of Mitchill's history of the quakes, she wrote that "it bespeaks a certain arrogance, to suppose that we have nothing left to learn from the voices of the past." Looking into the New Madrid quakes from the much higher pinnacle of geological knowledge that exists today, she wrote of the happy surprise of "discovering the extent to which contemporary observers displayed both credibility and insight. One is reminded that two hundred years ago, as now, there were some very smart people around."

That pinnacle of knowledge from which scientists view the world today might better be called the slope of a great mountain, since much about earthquakes and other dramatic features of the planet's metabolism remains unknown. But with the New Madrid earthquakes in 1811–1812, the scientific unraveling of earthquakes had begun, however haltingly, in the United States and elsewhere. The first true seismographs would soon be invented. Before the century ended, some seismologists would see their first "picture" of an earthquake. They would see on a piece of paper how an earthquake signs its name. And a hundred years after that milestone, scientists could talk without utter hubris about forecasting earthquakes, even such bizarre and enigmatic ones as the three big quakes of New Madrid.

PART TWO

THE EARTHQUAKE HUNTERS

The one who places the last stone and steps across to the terra firma of accomplished discovery gets all the credit. Only the initiated know and honor those whose patient integrity and devotion to exact observation have made the last step possible.

— Hans Zinsser

4

Myths, Maps, and Machines

THE NOTIONS REFERRED TO by Samuel Mitchill of New York of an alkaline or electrical cause of the New Madrid earthquakes were early quasi-scientific attempts to explain these titanic and often harmful events. They sound as far-fetched to modern ears, however, as the idea of some American Indians that earthquakes were motions of the giant turtle upon which the continent rested, or that of some Japanese at the time that pinned them on a giant catfish upon which the islands of Japan floated. Elsewhere, such motions were those of an ox, a hog, an elephant, even a mole. The Norse said it was the trickster god Loki, finally caught and restrained, writhing violently to duck venom dripping from a snake's fangs.[1]

Mythology is an often snooty word for someone else's religion, and before giggling at exotic beliefs, it is well to remember that a surprising number of people in the sophisticated Western countries like the United States still believe that earthquakes are acts of God that constitute a punishment for moral lapses by societies at large. It is well also to remember that science is driven by the same urgent need to explain the world as mythologies are, and it took a long time for science to emerge from myth (a word that also has taken on the meaning of untrue explanations).

One of the earliest people to bridge the continuum between myth and what we call reason was the Greek philosopher and geometer of the sixth century B.C., Thales of Miletus, who is credited with being the first thinker to try to explain nature via natural causes, as opposed to supernatural ones. All we know of Thales comes down as hearsay since he left no writings of any sort, but he evidently suggested a single material substratum for the universe — namely water. Water may have seemed lifelike since it could transform itself from liquid to solid or to vapor. In any event, Thales believed that the earth floated on a vast universal ocean and the storms there shook the land.[2]

Winds blowing through subterranean caves was another notion put forth by some ancient Greeks as the cause of earthquakes, and Aristotle not illogically saw a connection between volcanic eruptions and earthquakes, both perhaps caused by the movement of gases within the earth. He also suggested a connection between earthquakes and atmospheric events. The gases blowing through caves started fires just as thunderstorms cause lightning, and the fires sometimes burst through the surface rock (volcanoes) and caverns collapsed (earthquakes). Aristotle also noted that earthquakes could be classified by whether the ground motions were primarily vertical or horizontal, and whether vapor emerged from the ground, noting observantly that "places whose subsoil is poor are shaken more because of the large amount of wind they absorb." These Aristotelian precepts about earthquakes held sway in Western minds until well into the nineteenth century, though others were also suggested, such as alkalinity and electricity.

Then several new and productive avenues came into view in the study of earthquakes. One was to catalogue their effects. One of the more accurate accounts of earthquake effects was by the seventeenth-century physicist Robert Hooke, who is best known for being the first to state an inverse square law to explain planetary motion (and was bitter till death that Newton got more credit for it than he). He also came up with Hooke's law of elasticity, namely that stretching a solid body — such as one of wood or metal — is proportional to the force applied, a law that would play a role much later in the analysis of earthquakes, though Hooke never made the connection himself. In his 1668 *Discourse on Earthquakes,* Hooke noted that rising and sinking land were important effects of these terrestrial calamities.

Some sixty years later, the American polymath Benjamin Franklin noted that earthquakes often produced vast cracks and even chasms in the ground. These would later be called faults, but it would be a long time before anyone sorted out whether faults were the causes or merely the effects of earthquakes. For many years, both suggestions would seem equally plausible.

An earthquake question that challenged some geologists in the early nineteenth century was where they tended to occur. People tend to remember or keep track of earthquakes that happen locally or regionally, but if there was to be a science of earthquakes, getting an overall global picture of their occurrence was necessary. On the other hand, as Benjamin F. Howell, Jr., seismologist at Pennsylvania State University, has pointed out with a waggishness not always associated with the study of earthquakes, as long as quakes were considered God's punishments for people who had fallen into sin, then earthquakes by definition had to be common wherever people existed. Soon enough, however, it would become "apparent either that people of certain areas had fallen further from grace than others or that some other explanation for earthquakes was necessary."

The first systematic attempt to catalogue the world's quakes occurred when the German geologist Karl Ernst Adolph von Hoff began publishing lists for the years 1821 through 1832: in all, 2,225 earthquakes. Earlier, a handful of people had listed smaller numbers based presumably on inferior and generally local information. Other such lists emerged here and there, including one by the Frenchman Alexis Perry, who catalogued more than 21,000 earthquakes worldwide for the years 1843 to 1871. Perry hoped that via his cataloguing he could match variations in earthquakes with the seasons and the phases of the moon. It would fall to an Irish industrialist, however, to put the collection of earthquake data on a more useful scientific footing.

This was Robert Mallet of Dublin, who earned a degree from Trinity College of that city in 1830 and worked for the next thirty-odd years in his father's factory and engineering firm. Among the firm's more notable work was helping to create Ireland's railroad system, to install a 133-ton roof on St. George's Church in Dublin, and to build the Fastnet Rock lighthouse. The lighthouse, a cast-iron structure, came to be known as the Teardrop of Ireland, being the last sight many Irish emigrants had of their homeland as they sailed to America and other destinations. Mallet was also a self-made physicist as well as an engineer, later in life writing an important treatise on metallurgy and ballistics and the possibilities of developing huge, smooth-bore cannon. And he developed a profound and lasting interest in earthquakes in spite of the fact that Ireland is largely free of such events.

More discriminating than Alexis Perry, Mallet produced a list of earthquakes for a much longer period of time — from 1606 B.C. to A.D. 1850 — and included only 6,831 events. On the basis of the last 150 years of his list, a period in which he admitted his information was more complete, he suggested that about every eight months an earthquake occurred that was strong enough to destroy nearby villages. Such quakes he called "great." Lesser ones were "mean," and even milder ones were "minor."

From this information, he produced a map of earthquakes worldwide. For great quakes he inscribed a circle of 540 miles around what he took to be the epicenter (then presumed to be the site of the greatest damage). Mean quakes were given circles of 180 miles, minor ones 54 miles. Others before him had attempted such maps, but Mallet's was the first to provide a clear and largely accurate view of global seismicity.

On the map, belts of seismicity merged along the mountainous edges of the continents (where other observers had already noted that most volcanoes occur). What today we call the Ring of Fire — the highly seismic circle around the Pacific Ocean — showed clearly on Mallet's map, along with the belt that runs from the Mediterranean through the Himalayas into Indonesia. His map had less to show of those less common quakes that occur within continental boundaries, such as the New Madrid quakes.

Yet another of Mallet's earthquake enterprises was to invent the word "seismic." It was taken from the Greek for "shake." Yet, with all his interest in and work on earthquakes, Mallet had never experienced the shake of an actual earthquake himself until a relatively minor one in early November 1852, in Great Britain. Five years later, he had his main chance. In December 1857, an enormous quake struck the area of Naples not far from Vesuvius, the volcano that had buried Pompeii long before. So great was the quake that Neapolitan communications with the rest of the world collapsed for a week. As soon as Mallet heard about it, he applied for and received a grant from the Royal Society in London to travel to Naples. He stayed for two months, making what are widely taken to be the first truly scientific *field* studies of a great quake.

On the scene, Mallet set out to map and tabulate reports by those who had felt the quake, movements of the ground, and damage to buildings. He took note of the direction of cracks in walls and the patterns of rubble from fallen masonry, and drew lines on a map where damage was equally intense. He called these lines "isoseismal" lines, much like the isothermal lines on weather maps, which delineate the areas where the same ("iso") temperatures reign. From this information he could determine the direction of the source of the seismic waves, which is to say, the true epicenter of the quake. Not only was this a great improvement on previous attempts to locate a quake's epicenter, but the isoseismal lines suggested the rate at which shaking dropped off, giving him an idea of the relative size of the quake.

Mallet thought earthquakes were most likely the result of compression (of underlying rock) followed by a release of pressure. The association of many quakes with volcanoes was hardly lost on him. But beyond these observations Mallet made little effort to describe the engine or engines that drove earthquakes, while other scientists continued to speculate.

In 1849, the great German-French naturalist and explorer of South America Alexander von Humboldt published the first volume of his planned five-volume popular explanation of all that was known at the time of the physical universe. Called *Cosmos,* it was one of the most ambitious scientific works ever undertaken. In his chapter on earthquakes he took note of quakes where the earth had shaken for days at a time, writing that "I am only acquainted with the following cases in which shocks that have been felt almost every hour for months together have occurred far from any volcano." He cited three examples he knew of, including the quake that occurred "between New Madrid and Little Prairie, north [sic] of Cincinnati, in the United States of America, in December, 1811, as well as through the whole winter of 1812."

It is an interesting sidelight that in the same year, 1849, that Humboldt's great work began to appear, the founder of modern geology, the already knighted Sir Charles Lyell, published volume two of *A Second Visit to the United States of North America,* in which he recounted travels that included a visit to New

Madrid in March 1846, "where I intended to stay and make geological observations of the region. . . . So many of our American friends had tried to dissuade us from sojourning in so rude a place, that we were prepared for the worst." Upon arrival, Lyell and his wife found that the "only inn in New Madrid had been given up for want of custom" and they sought rooms elsewhere as night came on, eventually managing to cadge a room with a German baker in town. Setting out the next day to examine the still visible effects of the great quakes, Lyell drew attention to the contemporaneity of the terrible quakes in Venezuela, saying a bit inaccurately that Humboldt called the New Madrid shocks "the only examples on record, of the ground having quaked almost incessantly for three months, at a point so far remote from any volcano." Geologically, Sir Charles found little to add to the understanding of the New Madrid quakes except to suggest that the black bituminous material the quakes had thrown up here and there "was probably drifted down at a former period by the current of the Mississippi, from the coal fields farther north."

Here then is a lesson in the caution — or forgiveness — with which one should read the accounts of even the most prominent of scientists: Humboldt, the great geographer, put New Madrid north of Cincinnati, while Lyell, the great geologist, misrepresented Humboldt about earthquakes and volcanoes.[3]

Robert Mallet's work in the 1850s with his isoseismal lines in and around Naples was an early scientific effort to measure or otherwise get a firm bead on the intensity of earthquakes, a subject that would be a major preoccupation henceforth of those interested in the field, and Mallet became one of the first to ever actually *experiment* with such matters. He wanted to know how seismic waves fall off in intensity, and, rather than wait around passively for an earthquake to occur (which as noted was not likely to happen with any regularity in Great Britain or Ireland), he created his own. Setting off charges of gunpowder in the ground, he attempted to time the arrival of the shock waves at various distances from the explosions and as they traveled through both rock and sand. While the experiments were pioneering, and useful in that he found true what Aristotle had stated — that the waves go faster through rock than sand — his own efforts at making a seismograph were not really up to the task and he could learn no more than that.

Scientific instruments of a sort were made as early in history as the ancient Greek and Roman civilizations, but, like the Chinese "seismometer," they were used more to entertain or to illustrate theories than to make quantitative measurements. They were educational devices rather than tools for opening up nature's secrets, but later they evolved into actual tools for such activities as surveying and navigation. The world in the sixteenth century A.D. was still operating on the basis of phenomena that could be seen with the naked eye. In the seventeenth century, Galileo demonstrated that a tube fixed with lenses at either end could show humanity evidence never seen before (like the moons of Jupiter), and the modern age of scientific instrumentation began, and with it modern science. The microscope was not far behind, and as the late science historian Derek deSolla Price of Yale pointed out, by 1800 instrument-making had grown into a major industry. Forces previously unknown in everyday life — electric current, for example — would become fairly commonplace once an Italian named Volta invented the battery, and other such devices came into being. Driven for highly practical reasons to study earthquakes and volcanic eruptions, perhaps one day to be able to predict them, a handful of inventors produced seismoscopes, mostly based on the use of pendulums, as those of Jared Brooks of Louisville had been.

In 1839, Scotland was struck by a series of small shocks, originating near the town of Comrie, and they were felt for a number of years. The British Association for the Advancement of Science established a special committee to create instruments that could register and record the shocks. Of the designs that arose in answer to the society's call, one by a Scot, James Forbes, appeared to be the most promising. It was an *inverted* pendulum — in essence a vertical rod with a mass or bob toward the upper end. The rod was supported at the bottom on a strong vertical strip of steel while at the very top — and well above the bob — a pencil on the rod wrote a record of the movements on a paper-lined dome. The hope evidently was that the pencil would produce a straight line on the dome that would correspond to the sudden displacement of the earth. The straight line would be easy to distinguish from the ellipsoidal traces drawn when the pendulum oscillated around its new equilibrium position.

Six of these seismometers were set up around the town of Comrie, but they did not work as hoped. In one year they recorded only three of the sixty quakes felt there. What went wrong? It is not clear, but friction where the pencil met the paper is blamed, though the stiffness of the wire holding up the rod might have played a role as well. In any event, the Comrie quakes began to tail off, and after a few years the committee established to produce seismic instruments dissolved. Later models called for reducing the friction of pencil on paper by substituting a stylus that moved through light powder.

In 1851, as Robert Mallet began blowing up gunpowder to produce elastic waves in various kinds of surface rocks, he needed an instrument to spot them as they arrived at various distances from the explosion. For this he built a contraption that had light pass through a tube, reflecting the image of cross hairs located in the tube onto a containerful of mercury. Looking through an eleven-power magnifier, Mallet would see the reflected cross hairs blur or disappear when a slight shaking agitated the mercury. In the case of waves passing through granite, he got velocities of about 1,600 feet per second, which he al-

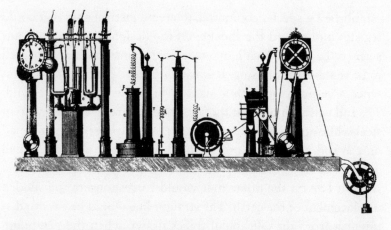

Figure 5. Luigi Palmieri's "sismografo elettro-magnetico" employed clocks, weights on springs, U-tubes of mercury, and other devices to measure horizontal and vertical motion, intensity, direction, and duration. Instruments with springs measured vertical motion. Derived from James Dewey and Perry Byerly, "The Early History of Seismology (to 1900)," *Seismological Society of America Bulletin,* 59, pp. 183–227, 1969.

ready knew from earlier experimentation to be too slow. Later seismologists using similar equipment would come closer to the real velocity for granite — 8,000 feet per second — suggesting that Mallet for whatever reasons — perhaps inadequacy of his measuring equipment — simply wasn't perceiving the earliest arriving waves.

Another problem to be overcome in creating a working seismo*graph,* an instrument that would record quakes, was hooking it up to a clock so that one would know when a shock occurred and how long it lasted. In 1856 an Italian, Luigi Palmieri, installed a "sismografo elettro-magnetico" at an observatory on Mount Vesuvius. A marvelous contraption, it consisted of several different kinds of seismoscopes, each intended to measure a different earthquake parameter — horizontal motion, vertical motion, intensity, direction, and (being hooked to a clock) duration. To detect vertical motion, for example, Palmieri suspended from a spiral spring a cone-shaped mass. It hung over a container of mercury. A slight motion made the tip of the cone touch the mercury, completing an electric circuit which in turn caused the clock to stop running, thus indicating the time of the shock, and also causing a drum holding recording paper to start moving and a pencil to be pressed lightly against it. A second clock regulated the speed of the paper. Both pendulums and U-tubes of mercury would sway, thus detecting horizontal motion. This pre–Rube Goldberg device worked. It recorded shocks on Mount Vesuvius for many years. Palmieri used it to record the frequent small shocks that, he found, often preceded and seemed to be predictive of volcanic eruptions and large earthquakes.

The Palmieri sismografo was eagerly imported by the Japanese, who dwell in one of the most seismically active places on the planet. Earthquakes were commonplace features of Japanese life, and of course caused a great deal of damage and suffering. Palmieri's instruments recorded earthquakes in Tokyo over a ten-year period beginning in 1875 — in all 565 quakes. Indeed, it would be in Japan that seismology would soon become what can be thought of as a mature and modern science.

The Japanese Connection

Beginning in 1865, the government of the Emperor Meiji determined that the xenophobic isolation and virtually medieval ways of Japan had to end. The nation needed to catch up with the industrial might of the West, and had to do so rapidly. To that end, the government began importing what they called *oyatoi-gaikokujin*, literally, honorable foreign menials. These were specialists in various areas, economic as well as technical and scientific, brought in to train Japanese nationals. In 1875, the Japanese issued an invitation to an almost manically energetic young English geologist named John Milne.[4] In his twenties, Milne had already done geological fieldwork in, among other places, Iceland, Arabia, and Newfoundland, specializing in mining but fascinated by virtually every aspect of geology. The Japanese offered him a professorship at Tokyo's Imperial College of Engineering, which at that time was the largest technical college in the world. There he would teach both mining engineering and geology.

Being given to severe seasickness, Milne elected to travel overland eight thousand miles to Japan, an arduous trip that took him some six adventurous months, arriving in Tokyo in March 1876. On his first night there a sizable earthquake shook the ground — the first Milne had ever experienced — and he was hooked. While most of the honorable foreign menials stayed in Japan for relatively brief periods — a few years — Milne stayed for twenty years, married a Japanese woman, and became the undisputed father of modern seismology.

His professorial job teaching mining engineering and geology included wide travel throughout the Japanese islands to advise the government on mineral extraction, and in the course of these travels he also catalogued and studied virtually all of the Japanese volcanoes. This involved clambering around the craters of active volcanoes at considerable risk, attempting to measure their diameters, for example, even as they gave off huge clouds of vapor that made seeing difficult. It was commonly held in Japan that most of the country's many earthquakes were brought about by volcanic activity. Milne disabused his colleagues of this, pointing out that in "the centre of Japan there are mountainous districts where active volcanos are numerous, yet this area is singularly free from earthquakes." Milne also found time to dabble in archaeology, excavating kitchen middens and opining on such matters as the then mysterious origin of the Japanese people themselves. Throughout his years in Japan, however, great and small movements of the earth would not let John Milne alone.

"In April," he wrote, "the month following my arrival, there were no less than ten shocks recorded," and before long he had joined with others of the faculty — both English and Japanese — in trying to unravel their mysteries. One key, Milne knew and persuaded his colleagues, was better instrumentation. Another key was to involve more people in the pursuit. At the time there was no such thing as a seismological science, no group of scholars devoted to the study. It was at best the interest of a few isolated people whose training lay elsewhere. He proposed the formation of the Japanese Seismological Society, the first such group anywhere, and it held its first meeting in April 1880. The importance of such a society cannot be overestimated. Where before seismology was done by lone hobbyists working with considerable ingenuity, now it was — in Japan at least — a large group of like-minded scholars focusing on a single large problem, with all the synergy possible in such a group.

Milne, who continued to supply much of the energy behind the society, experimented with all sorts of measuring devices, concluding (as had Jared Brooks in Louisville long before) that the principle of the simple pendulum was best suited for a proper seismograph. He filled much of his house with pendulums of varying lengths (again, as had Jared Brooks), but his were more ambitious to say the least: they ranged from one to thirty-six feet long, the better to mark the magnitudes of the frequent quakes. Milne worked particularly closely with two other engineers who were in Japan for a few years — James Alfred Ewing and Thomas Gray. Ewing took the lead in producing a seismograph that employed a horizontal pendulum that could swing back and forth not unlike a "garden gate," as historian James Dewey and Perry Byerly have written. This device produced a vertical line on paper until a tremor made a sideways trace. It recorded a small quake in 1880, providing the first image of earth-

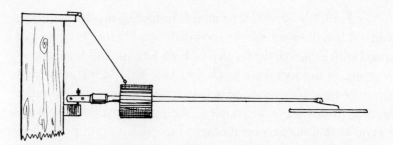

Figure 6. One of Ewing and Gray's seismographs employing a horizontal pendulum that swings in the manner of a garden gate. The scriber at the far end of the pendulum records the motion on paper. Derived from James Dewey and Perry Byerly, "The Early History of Seismology (to 1900)," *Seismological Society of America Bulletin*, 59, pp. 183–227, 1969.

quake motion over time. With this and records of a few subsequent quakes, Ewing took note of some important and hitherto unknown features of earthquakes.

> (1) The very gradual beginning and ending of the disturbance. In none of the observations did the maximum motion occur until several complete oscillations had taken place. (2) The irregularity of the motion. The successive undulations are both widely different in extent and in periodic time. (3) The large number of undulations in a single earthquake, and the continuous character of the shock. (4) The extreme minuteness of the motion of the Earth's surface.

For the first time scientists could look at a real drawing, a model, of earthquake motion — and it was different than what anyone had earlier supposed. (Robert Mallet, for one, thought an earthquake consisted only of a single longitudinal pulse.) With this picture in mind of what seismographs should be recording, it would be possible to build yet better, more accurate and responsive instruments.

Others were not far behind. Within a year or two Gray and

Milne developed an instrument for detecting vertical ground motion; Ewing designed a "duplex-pendulum" seismometer that employed a common pendulum and an inverted one (with the bob toward the top), and instruments of this sort soon found their way across the Pacific and were placed in spots in northern California and Nevada. Meanwhile Milne (following Mallet's footsteps) took up using dynamite to propagate earthquake-like waves in various topographies and through various media, using Ewing's and Gray's seismographs to find out their speed and other parameters. He also used such man-made "quakes" to test their effects on houses of differing sizes, materials, and designs so that he could advise the Japanese government on the best way to build houses that could withstand the regular assault from Japan's seismic instability.

Ewing and Gray went home shortly after the above-mentioned seismographs were produced, and Milne continued his efforts in league with Japanese colleagues. As the English and other foreign members of the Japanese Seismological Society left, the society's efforts became somewhat attenuated, its meetings less well attended.

Then in 1891 a massive quake struck central Japan — "as terrible," Milne wrote, "a shaking as has ever been recorded in the history of seismology." The devastation was widespread and savage, with railways contorted, cities and towns shaken down, and mountains slipped downward, damming up whole valleys. Almost ten thousand deaths were reported. The emperor immediately created the Imperial Earthquake Investigation Committee, with Milne its only foreigner among the geologists, architects, engineers, and others commanded to determine if earthquakes could be predicted and if buildings could be made safer.

Milne produced a fifty-question questionnaire, and some ten thousand copies were distributed to citizens and filled out. These and hundreds of seismograms from a huge array of seismographs spread around the country, along with newspaper accounts and photographs, provided the most detailed account of an earthquake and its effects ever compiled up to that time. Milne and others attributed the quake to a long fault that stretched through central Japan, but what caused the fault to move was another question altogether. Milne was little interested in such theorizing, preferring the tasks of measurement and information exchange.

By the 1890s, Japan led the world in seismological research and, but for Milne, it was all in the hands of Japanese scientists. Of these, Milne's special protégé was Fusakichi Omori, and the two collaborated on many research projects, with Omori becoming an internationally recognized specialist in the study of aftershocks. At the same time, seismology was becoming international in scope — and in more ways than one.

Global Reach

Almost immediately, Europeans began to perceive the successes in seismograph design in Japan, and were importing the instruments and making minor improvements on them. The most exciting seismological news came, however, from a man who was trying to determine the effect of the moon on gravity. A young German, Ernst von Rebeur-Paschwitz, built astronomical instruments to measure extremely minute changes. Eliminating friction as much as possible was crucial, and he became the first to employ photographic recording for continuous observations. He employed light reflected through a lens to a rotating drum covered with photographic paper. Thus friction was completely absent from the recording apparatus.

Von Rebeur located two of these devices in Potsdam and Wilhelmshaven and on April 17, 1889, recorded a small tremor.

He discovered later that it had been a trace record of a very large earthquake that had been felt about an hour earlier — *in Japan.* When Milne, still in Japan, heard of this he began his own studies of what came to be known as "teleseisms" (literally, distant shakings) and developed an instrument that would record them all the more accurately. His too was a horizontal pendulum coupled to a photographic recording system in which photographic paper moved continuously.

In 1895, Milne and his wife moved to England, where he established his own seismological observatory in a small town, Shide, on the Isle of Wight. There he devoted the rest of his life to creating ever more sensitive seismographs for the recording and analysis of teleseisms. Since Milne had realized that with a global array of proper instruments one would be able to record virtually any earthquake anywhere in the world, he persuaded the British Association for the Advancement of Science as well as the government and private individuals to finance the installation of a worldwide network of his seismographs. With the resulting information being constantly fed back to the Isle of Wight, he produced the *Shide Circulars,* which he shared with the rest of the seismological world.

By 1902, Milne had arranged for some sixty seismological observatories in forty countries (mostly in the British Empire) which could monitor quakes in most parts of the world. One of his instruments was taken to Antarctica by Sir Robert Scott in the *Discovery* and set up at base camp. In the same period, with impetus from German seismologists, an international committee for earthquake research got under way with which Milne and his array cooperated — though a bit hesitantly, knowing that other German activities at the time were coming under international criticism and, if they were to lead to war, would upset such a system.

In addition to all of the above, Milne continued to produce catalogues of current and historical earthquakes, looked into such phenomena as earthquake luminosity, and, perhaps most significantly, determined a satisfactory way of locating the epicenters of distant quakes. Milne spent his latter years pottering around in his seismological lab, producing little that was new. In July 1913, he died, and the emperor of Japan sent his brother-in-law to represent him at the funeral. In addition to the many condolences his wife received from Milne's scientific peers, she received a letter from the Japanese ambassador to Great Britain that said, "It is not only an irreparable loss to this country and the scientific world but also to Japan where his name will never be forgotten."

The Waves of Assam

In 1897, another British geologist found himself with an important foreign appointment. Richard D. Oldham was made director of the Geological Survey of India, a post he held for six years. During his first year there one of the most powerful earthquakes ever recorded struck Assam, killing more than a thousand people. He was responsible for what became the famous study of the Assam earthquake, which, among other things, assembled every record available from the seismographic observatories in Europe and elsewhere. From poring over these teleseismic records, Oldham determined that there were several different kinds of waves emanating from the great quake. One set of waves arrived at a given seismograph ahead of another set of waves. These have come to be known as P waves (for primary) and S waves (for secondary). P and S waves are called body waves since they penetrate the body of the planet. There are also surface waves that emanate from the epicenter of an earthquake — often the most damaging — and these had already been noted by other researchers.

To understand how the basic kinds of waves work, imagine yourself throwing a rock into a pond. The impact sends concentric waves spreading out across the surface of the water, but the pressure of the rock hitting the water also sends out waves throughout the water. They compress the water slightly as they pass through it.

Something quite similar occurs when shock waves go through rock. As the P waves, those first ones to arrive at a seismographic observatory, move through the earth — be it rock, liquid magma, or even water — the molecules tend to compress ever so slightly and then dilate — much as the air compresses and dilates when sound waves go through it. The molecules, in effect, jiggle back and forth along the direction of the P wave. The harder the substance, the faster the P waves go. Indeed, they behave so much like sound waves that if they escape through the surface into the air, they may sometimes be heard.

S waves on the other hand tend to shear the molecules of material they pass through, pushing them up and down or from side to side. Essentially, they force the rock to twist sideways and then spring back. True liquid, however, does not deform, so S waves do not go through liquid materials.

P waves, arriving first, are something like a sonic boom rattling the windows. S waves come along later, more slowly, more violent, shaking the ground. Two kinds of surface waves follow along after these two body waves have passed by. Named for their discoverers, they are Love waves and Rayleigh waves. The motion of Love waves is like that of S waves — side to side — causing horizontal shaking of the ground (and any structure upon it). Rayleigh waves on the other hand are more like the rolling waves of the ocean, lifting the surface and dropping it at the same time, pushing it in the direction the wave is headed. Learning about these various waves was not only the key to understanding the action of earthquakes but also the key to peering into the innards of the planet. And both of these bits of progress would lead to a revolution in geological understanding that would shed light on the tantalizing enigma of the New Madrid quakes.

At the time — the turn of the twentieth century — teleseismic waves had only recently been discovered by Von Rebeur, and like Milne, Oldham concentrated on them, soon doing what is generally thought of as one of the most brilliant pieces of detective work in the history of seismology.

"X-raying" the Earth

"The seismograph," Richard D. Oldham wrote in 1906, "recording the unfelt motion of distant earthquakes, enables us to see into the Earth and determine its nature with as great a certainty as if we could drive a tunnel through it." Oldham had to be elated when he wrote those lines, since he had made a fantastic discovery, but his certainty was perhaps a bit premature. What he had just done, and what no one had ever done before (except in fantasies like those of Jules Verne), was to get a first glimpse of the earth's actual interior anatomy.

Up until that point, seismologists had been able to gain a general sense of what the earth was made of a few miles below the surface. But was the stuff of the earth's center solid, gaseous, fluid? A combination? No one knew. Was the stuff deep down in the earth's interior more like raisin bread, with little chunks of stuff embedded in some other stuff uniformly throughout, or was it more like an onion with many concentric layers? No one knew. Aside from impossibly prodigious digging, the only way,

Oldham realized, to look into the earth's interior was by using those teleseismic earthquake waves — P and S. So, with all the seismic records of earthquakes he could find, he began his investigation.

One common way of explaining what goes on in the earth with such waves is to imagine the earth as a giant glass ball that acts as a lens. An earthquake on the surface is like turning on a light bulb — it sends waves in every direction into the crystal ball (and outside it as well, which does not happen with seismic waves in the earth). A wave going straight through the glass ball to the opposite point from the light bulb would have gone, in other words, to a point 180 degrees around the ball, or halfway. Oldham found, however, that P and S waves did not travel straight through to the other side of the world from a given quake, but only partway — never more than a point that was 110 degrees from ground zero. He hypothesized that the best explanation for this behavior was the existence of a large central core to the earth through which the S waves could not penetrate. Not long afterward, in 1914, German geophysicist Beno Gutenberg refined this finding. He noted that P waves, beyond 105 degrees, slowed down a great deal, arriving at a seismograph on the surface something like five minutes after they should. It appeared that some P waves had actually bounced off the boundary of the earth's central core; and knowledge of the angle involved as well as the circumference of the earth, along with fairly straightforward geometry, led him to the estimate that the boundary of the core lay 1,740 miles below the surface.

With the onset of the twentieth century, then, the geological world was well poised for a century of fantastic discoveries about the earth, its innards, its composition, its long, long history, and its frequent violent seizures. By the end of the twentieth century seismologists would be able to measure and record disturbances in the earth so small that the early seismological tinkerers would have been unable to imagine them. Today's instruments can sense a displacement of the ground as small as the amount of space between atoms. By the end of the twentieth century we would have not just a plausible idea of what causes earthquakes — though not all of them — but, some would suggest, a chance of finding out how to predict them.

5

Finding Faults

EVEN IN THESE DAYS of virtual reality, to learn heart surgery
it is a good idea to work on an actual living human being. Just so,
learning about earthquakes is often best done when one has just
occurred. Between 1872 and the turn of the century, several studies of big quakes led geologists to make the leap to the belief that
it was slippage along faults that actually caused some earthquakes — not the other way around.

Even as early as Robert Mallet's researches in the 1850s, earthquakes were perceived as the events that over time built mountain belts and other great terrestrial structures. This process came
to be called tectonics. Even earlier, in 1837, Sir Charles Lyell had
confirmed that great earthquakes and large changes in the earth's
surface — faults — often occur at the same time. What was not
known was which caused what. It seemed likely, however, that
whatever enigmatic forces were responsible for earthquakes, it
was the explosive force of the quake that brought about the tear
in the fabric of the earth's surface. Sorting this matter out was, of
course, crucial to the developing understanding of earthquakes,
and would eventually lead seismologists to some important in-

sights into the New Madrid quakes — though not until many decades had passed.

The earliest such study was made by a Japanese seismologist and geologist, Bunjiro Koto, after a tremendous quake struck Japan in 1871. He found a sixty-mile-long fault whose two sides had shifted about fifteen feet. He was so deeply impressed by the extent of this fault that he concluded it could not have been the result of some underground explosion or the movement of magma beneath the surface (the commonly assumed causes of quakes at the time). Instead, it had to have been the huge slippage of the fault itself that had caused the quake. Two years later in California more light on this matter would be shed when a quake we now think may have been almost 8.0 in magnitude struck the western slope of the Sierra Nevada, in a place called Owens Valley.[1]

The Owens Valley Quake

Even today, not many people live in the Owens Valley, which runs north and south between the Sierra Nevada and the Panamint Mountains that seal Death Valley off from the rest of California. But when the big quake hit in 1872, literally obliterating towns like Lone Pine, only some sixty people are known to have perished. The initial shock was the greatest, creating a wave in Owens Lake that sent fish flying into the air. Aftershocks continued for two months. Along one hundred miles of what is today called the Owens Valley Fault the ground was displaced more than twenty feet in places.

One of the eyewitnesses of the quake was John Muir, the great naturalist and writer, who was asleep in his cabin in the forest above Owens Valley when the quake struck. Once outside, he had to hide behind a large tree to avoid the boulders loosened in the quake and now rolling down the mountain all around him. (Muir was the sort who would climb a tree in a gale, the better to feel the storm.) Once things had calmed down a bit, he put water in a bucket on a table, his own primitive seismoscope.

A more important (and more scientific) presence at the site of the quake came a little later — a soft-spoken geologist from the United States Geological Survey, Grove Karl Gilbert. Gilbert had been on several of the great scientific expeditions into the West, including one with John Wesley Powell, the explorer of the Colorado River and later head of the largest government-sponsored scientific agency on earth at the time, the Geological Survey. Gilbert continued working in the West, reaching the Owens Valley a few days after the quake. Among other things he noticed was that the mountains bounding the west side of the valley had moved upward and away from the valley floor. In other words, it appeared that the valley itself had widened eastward. This was similar to faulting that he had seen elsewhere in the West, especially in Utah, where terraces of rock showed previous uplifts that he thought had to have come about when similar vertical faulting occurred in the recent geological past. This confirmed the work of Bunjiro Koto — that it is the faults that cause earthquakes and not vice versa — and also laid the groundwork for the subfield of paleoseismology, whereby geologists can establish the age and sometimes the magnitude of prehistoric quakes.

Gilbert would go on to serve as head geologist of the United States Geological Survey (USGS), make valuable studies of the lacustrine nature of Lake Bonneville (part of a onetime inland lake of great size), and explain the formation of the Basin and Range country. No one is perfect, it seems, and Gilbert would take the wrong side in a debate over exactly what had caused the enormous crater that lies between Winslow and Flagstaff, Arizona, a looming and eerie presence on the dry high desert, and

today a privately owned and unselfconsciously tacky tourist at-
traction called Meteor Crater.

At this time, the late nineteenth century, the crater was called
Coon Butte, a circular feature some 4,000 feet across with walls
some 160 feet high and a depression some 600 feet deep. Odd
pieces of iron would occasionally show up on the ground nearby,
and Gilbert, in 1891, visited the place assuming that, were it the
impact crater of a meteorite, there would be a huge magnetic difference between the crater and the surrounding desert thanks to
a massive piece of metal in the crater. Magnetic surveys, however,
showed no such anomaly, so Gilbert opined that the crater was
the result of a massive explosion of steam from under the earth.
Later a mining engineer named Barringer heard about the place,
believed it had to be from a meteorite impact, bought it, and
hoped to make a fortune mining what had to be the huge chunk
of valuable metal under there. He drilled and drilled and lost his
shirt. Later, it was determined that the calculations people had
made about the speed and size of the meteorite were wrong. A
slow-moving meteorite would have to have been immense to create so large a crater, but one moving at the speed of a bullet
would have been small enough to have been largely vaporized at
impact, leaving no magnetic anomaly. The Geological Survey
continued to claim a volcanic origin until many years later, and
Gilbert himself never commented but probably believed that the
impact theory was after all correct.

Gilbert once wrote that it is "the natural and legitimate ambition of a properly constituted geologist to see a glacier, witness an
eruption and feel an earthquake." All one needs to do, he wrote,
is go to a glacier passively waiting, exercise a little alacrity and
get to a volcano in mid-eruption, but "the earthquake, unher-
alded and brief, may elude him through his entire lifetime." The
latter part of Gilbert's natural and legitimate ambition would be
realized in 1906.

San Francisco

A few minutes past five o'clock in the morning on April 18, 1906, some 6 miles down in the earth below the Golden Gate in San Francisco, a tiny fracture occurred in the rock and expanded outward in various directions, soon reaching a speed of 5,600 miles per hour, about ten times the speed of a cruising jetliner. It reached the surface in a matter of seconds and raced in both directions along the San Andreas Fault. On the surface above the original fracture — the epicenter — the ground on either side of the fault moved to the right about twelve feet. To the southeast the fracture disappeared from the surface after racing down the San Francisco peninsula some 65 miles. In the opposite direction, the fracturing tore northwestward, the displacement of the ground increasing to about twenty feet, sending out shock waves that damaged numerous towns of the northern Coast Ranges, reaching as far as Cape Mendocino, some 200 miles from the epicenter. As many as three thousand people may have been killed. The damage to structures in San Francisco itself was immense, but the major cause of destruction was the sixty separate fires that broke out, turning the city into a conflagration that lasted three days. A huge fraction of the city was destroyed — in all some forty-two thousand buildings.

No earthquake in American history has caused more deaths and more structural damage, even though it was at least ten times less powerful than the three quakes that struck New Madrid in 1811 and 1812. In perhaps no other earthquake have the agonizing tales of human loss been more fully documented — and oft told — than in the San Francisco quake. Perhaps the most famous story is one of the least awful, indeed even amusing. It tells of the great tenor Enrico Caruso, who was in town to sing *Carmen,* rushing out of his hotel onto the sidewalk, with his neck wrapped in a towel and clutching a signed photograph of Teddy

Roosevelt. There he was heard to cry at the top of his estimable lungs, "Give me Vesuvius!" Thereafter, checking out of the hotel, he vowed never to return to San Francisco, and he never did. Of more lasting importance, though, up to this time perhaps no other earthquake except the one in Japan that brought about the Imperial Earthquake Investigation Committee had been more fully investigated by scientists in the days following the catastrophe.

Gilbert, who was in Berkeley across the bay at the time, woke up on April 18 to "a tumult of noises and motions" and "it was with unalloyed pleasure that I became aware that a vigorous earthquake was in progress." The redwood house creaked, furniture rattled, and electric lamps swung, but the tremors were relatively "trivial," and it was not until two hours later that he learned that San Francisco was in flames. He was one of numerous distinguished geologists who were asked by the California governor three days later to form the California Earthquake Investigation Commission. Within a few days some twenty scientists were mounting a coordinated analysis of the disaster, as well as other groups that included architects and engineers.

The Japanese promptly sent its own commission, under the leadership of John Milne's old colleague, the aftershock expert Fusakichi Omori, to assist the Americans. The California commission was headed by Andrew Lawson of the State University in Berkeley, and the two detailed scientific reports that were later issued bear his name. So great was the detail in these reports that they are still being mined by seismologists today for new clues. Prior to their publication, however, Gilbert produced an interim report in which he pointed out, as if iterating what every geologist then knew, that "the displacements of rock masses are the primary and important phenomena: the faults are incidental phenomena, of great value as indices of the displacements; and the earthquakes are of the nature of symptoms, serving to direct attention to the fact that the great earth forces have not ceased to act." By contrast, he said, faulting far below the surface of the earth, as in the New Madrid quakes and that in Charleston, may go unnoticed except as the earthquake itself.

Gilbert included some striking anecdotes about the San Francisco quake. He noted that among other disruptions like road-blocking landslides, changes in the water table here and there, and eruptions of sand, many redwoods that had managed to grow right on "the line of the rift were split from the ground upward." Early findings indicated that the parts of the city that experienced the most appalling structural damage were those on "built" land — that is, land that had been reclaimed from marshes and the bay by filling in with sand and rock. As little as five miles east or west of the fault, the damage was much diminished, the quake not nearly so severe. Recall that Gilbert, just across the bay when the quake hit, felt it but had no idea that it was so severe only a few miles away.

One of the important puzzles about such faulting, Gilbert noted, was whether the ground on both sides of the fault moved — with the western side lurching northward and the eastern side lurching to the south, or was it only one side that moved? A means of resolving the issue was at hand: the United States Coast and Geodetic Survey had previously set up a series of surveying points along the coast and elsewhere to better refine topographic and hydrological maps for the state. Resurveying would provide the answer. A further check was possible at an observatory some twenty-five miles northeast in Ukiah. The observatory was one of an international series of observatories designed to determine variations in the position of the earth's axis of rotation. So refined was this instrumentation that one could tell if the observatory had moved several hundredths of a second to the southeast, a second being in this case one thirty-sixth of a degree of arc.

As it turned out, it was mostly the land on the western side of the fault that lurched so minutely northward. We know today, of course, that as this sliver of the continent continues its motion over the millennia, in a mere eight million years, Los Angeles, which is now west of the San Andreas Fault, and San Francisco on its east will be twin cities like Minneapolis and St. Paul.

The second volume of the Lawson report was, in fact, written by Henry Fielding Reid of Johns Hopkins University and one of the original members of the California earthquake commission. Published in 1910 by the Carnegie Foundation, which also funded the commission's work, it described the mechanics of the great quake and enunciated Reid's theory of "elastic rebound," which he took to be the immediate cause of this quake and others like it. While it is difficult in a certain way to imagine rocks behaving in a similar manner to something as flexible and mundane as a rubber band, one can mimic something of Reid's theory of elasticity by stretching a rubber band to its limit and then cutting it. All the pent-up energy from the stretching is released with a sudden snap. Similarly, friction locks the ground together on either side of a fault and the force of one side straining to move against the other results in the ground bending. Finally, the locking effect of friction is overwhelmed at the weakest point in the rock and the elastic energy that has built up is suddenly released, as the two sides rebound (just as a rubber band regains its original shape once you stop stretching it). The energy that was stored up, upon release, explodes outward, in part as heat and mostly as elastic waves. It is these waves that are the earthquake.

Reid's elastic rebound theory was accepted without much by way of professional demur and is indeed the basis of much of today's thinking about the immediate (or, to use a fancy scientific term, *proximate*) causes of most earthquakes. On the other hand, at the time, the only notion that could be adduced as to the *ultimate* cause — that is, what was making the ground move against itself — was a reference to the mysterious, unceasing "earth forces" that build mountains, mentioned by Gilbert in his report and by others. On the topic of ultimate causes, Reid had nothing to say.

The great quake of 1906 established California as the center of earthquake science, and for good reason. It was the most seismically active part of the United States at the time. That the 1906 quake had occurred where it did (on the San Andreas Fault) was no surprise to anyone with any geological knowledge and many with none, for earthquakes of varying intensity were common features of the region, just as they were in Japan. The San Andreas Fault is certainly the most famous fault in the world, and it would soon become the most studied such fracture anywhere. In fact, it is not a single fault running some five hundred miles roughly northwest to southeast through California but a large series of faults that are braided together. The other faults appear to have been splayed off from the main fault, or have come about in compensation for the seizures of the main fault. Of course, earthquakes have been occurring along this fault and its branches and twigs from time long before memory. The 1906 quake was something of a reprise of one that hit approximately the same area in June 1838, a quake of comparable power. Another major shock struck the same area (with the epicenter at Hayward) in October 1868, causing property damage of more than $300,000 in the settlement of San Francisco. These two quakes have been estimated at about 7.0 in magnitude.

After the 1906 earthquake, a tremendous amount of scholarly effort and technological genius was expended to come to grips with the mechanisms of earthquake faults. They were sought out, mapped, measured, probed, and, of course, classified. Seismologists learned they come in two basic types. In one, only horizontal displacement occurs: this is a strike-slip fault. There are two kinds of strike-slip faulting — judged from where

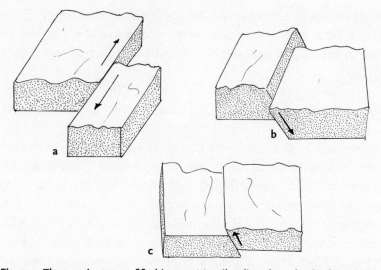

Figure 7. Three major types of faulting are (a) strike-slip, where the displacement is sideways, (b) normal, where one side drops below the other, and (c) thrust, where one side shoves upward above the other.

you stand looking across the fault. If the movement was from your right to your left, then left-lateral faulting is what occurred — and vice versa.

Two kinds of vertical — or dip-slip — faulting occur as well. When the side of the fault that overhangs the fault line slips diagonally downward relative to the other side, it is called a normal fault. When the side that hangs over the fault line moves upward, it is called a reverse or thrust fault. Of course, what actually happens in a quake is not always so simple. A fault can show strike-slip and dip-slip faulting in the same place.[2]

While such matters were becoming clearer and clearer, and could increasingly be characterized in numbers and formulas, many questions about faults remained unanswered, even in some cases not asked. The world of seismology was full of enigmas, and still is, but no earthquakes remained more enigmatic than the ones that had taken place a century earlier so far away from

the great seismic boulevard of the San Andreas Fault — the New Madrid quakes.

A No-Fault Quake?

One would guess from his most noted works that Myron L. Fuller was no man of the mountains. One of his publications for the United States Geological Survey was an explanation of the geology of Long Island, a place not known for a landscape rising high above the surrounding sea. He also worked on the nature of parabolic dunes, those mysterious sickle-shaped features of some deserts. And in 1912, the Geological Survey published his (for the time) exhaustive study, *The New Madrid Earthquake.*

Fuller had been drawn to the area in 1904 by Professor Edward M. Shepard, who was interested in certain matters to do with artesian water sources there in relation to the 1811–1812 quakes. Fuller traveled with him several times, by dugout, horseback, train, and wagon. Over the years Fuller covered most of the ground, checking out the old stories of eyewitnesses and matching them to the record apparent on the ground. Early on in his treatise he scoffed at the reports of flashes of light some reported during the quakes, suggesting that they "doubtless" arose from the generally held notion at the time that earthquakes were caused by volcanoes. Later on in the work he softened a bit, calling it "improbable" that they were all imaginary. Some might well have been thunderstorms on the horizon. He also cited unusual light shows occurring after a Mexican earthquake in 1907. It seems that various sea captains in the tropics saw strong glows resembling the northern lights: "It is not improbable," Fuller wrote, "that similar magnetic manifestations were associated with the New Madrid shock." Today we know that the auroras of the north are caused by ionized particles that are shot out by the sun — and called the solar wind — and interact with the earth's magnetic field, but surely the New Madrid light effects were not the result of the solar wind but something else altogether, if they were in fact actually seen.

Fuller went on to consider the science-minded adventurer Louis Bringier's explanation of a sinking surface and sudden ejections of sand, water, and other materials. Bringier thought they came about when underground cavities collapsed. Fuller suggested instead that they resulted from the abrupt opening and then closing of fissures in the land. The sulfurous stench that many reported as being associated with these geyserlike ejections, Fuller guessed, came from rotting organic matter in the alluvial soils that was thrown up to the surface during the shocks. The sudden blackness of the night was surely the result of the dust thrown up. The warm water people felt was probably just that — groundwater of about fifty degrees thrown up into air that was probably closer to freezing temperature.

One of the more astounding tales that by Fuller's time — a century after the fact — had entered the folklore was that the shocks were so great that the mighty Mississippi had flowed north. Timothy Flint had mentioned "a reflux of its waves," and others had seen much the same thing. Fuller agreed. There was no reason to doubt, he said, that great upheavals of the riverbed caused the water to be thrown back, parting like the Red Sea, with waves thrashing back and forth across the river, but also running retrograde for moments, especially in shallow water. And, one might add, at a time like that, a moment of retrograde flow would seem of far greater duration. All in all, while Fuller thought that "calm observation and accurate recording" of such a dangerous event was not even improbable but impossible, he nevertheless could conclude that "most of the reported phenomena have been verified by the recent investigations." This was chiefly because of the fortuitous presence in the area when the shocks occurred of "a number of scientists or men of education."

But if earthquakes were a matter of faults acting up, the New Madrid quakes were not about to yield any real secrets to the observant Myron Fuller. The place was covered with ridges and fissures — large stretches looked like Bunyanesque corduroy seen from the sky, the fissures presumably being the result of the wavelike heaving of the earth's surface, the land being described by many eyewitnesses as not unlike the ocean. And here and there, while much of the land was sinking, domes had been thrown up. Perhaps the exemplary one of these is the Tiptonville Dome, which runs south from New Madrid about fifteen miles and fetches up across the river about five miles from Reelfoot Lake. The upper portion of the dome consisted of the alluvial soils of the river bottom, now lifted some twenty feet above the Mississippi.

But few actual faults were to be found. The two locations of abrupt subsidence in the river itself, causing the two waterfalls above and below New Madrid, qualified as faults, as did a few other places where the land subsided or two sides of a fissure remained at different heights. But for the most part, the ground truth of the matter put the New Madrid quakes in a largely undefinable category all by themselves.

Fuller found it unreasonable to imagine that an earthquake that was centered in the alluvial part of the Mississippi basin could ramify through the hard rock of the Appalachians and be felt to their east. He concluded that there were, therefore, multiple epicenters that moved from the southwest toward the northeast in the general New Madrid region, but later shocks that had been felt as vertical oscillations in such places as Detroit had to have come from epicenters farther north. The ultimate cause of the New Madrid quakes, Fuller opined, was some sort of major disturbance beneath the embayment of the Mississippi River. A great subsidence caused by faulting deep under the surface, with the lands essentially falling inward, could explain most of the features of the land that came about during the quake. But he arrived at a different conclusion. He suggested that over the eons, while the river eroded the land downward, forces operating in the depths were perhaps forcing the same land upward, the two processes achieving short periods (geologically speaking) of equilibrium. "It is even possible," Fuller wrote, "that the shocks of 1811–12 were incidents of an uplift rather than of depression."

In other words, while seismologists around the world were beginning to perceive many of the mechanisms of these dreadful earth seizures, the causes of the New Madrid quakes remained utterly enigmatic. As for the ominous question of the future, Fuller wrote, "prediction as to when a repetition of the shocks will take place is futile."

Of course, in Fuller's time near the beginning of the twentieth century there was a great deal more to learn about earthquakes before prediction could be much more than a dream of the future. For one thing, seismologists needed some way of measuring how bad quakes were, how big or how small, so they could be compared. And this meant numbers. Today every earthquake that is recorded is given a number, and it is from the numbers that we — the public — believe we know how bad a bad quake is.

6

Intensity, Magnitude, and Stars

IF ANY NAME is associated in the public mind with earthquakes it is Charles Richter. With the news of any earthquake, the press (and the public) first want to know how it rates on the Richter scale. Most of us know that anything over 6 on the Richter scale is likely to be a lot of trouble and an 8 is an utter calamity. Beyond that, for most people, the numbers are otherwise meaningless.

Indeed, Richter used to say that some people thought it was an actual physical scale like a bathroom scale, and asked to see it. But it is simply a mathematical scale. What exactly does it measure? It measures what is called magnitude, which, in stars, means brightness — to know a star's magnitude is to be able to compare its brightness to that of some other star. In fact, when Richter developed his scale, he was asked exactly what aspect of an earthquake it measured, and Richter, an amateur astronomer, imported the word "magnitude." A number given by the Richter scale is based on the greatest amplitude of waves reaching a seismograph, or several seismographs. It is designed to permit a comparison between earthquakes. What the Richter scale does not measure is an earthquake's intensity.

The intensity of an earthquake is a measure of an earthquake's actual effects — in essence the actual damage it does. It is based on what is really subjective judgment. And of course an earthquake's intensity can vary widely even over a small area — the San Francisco quake struck with terrific intensity near the Golden Gate but hardly at all near Berkeley. Intensity is actually the scale that provides the kind of information that the press and the public really want to know: How bad was it?

As we saw earlier, Jared Brooks, patiently sitting in his house in Louisville watching his pendulums and spring in the winter of 1811–1812, was one of the first people to develop a fairly detailed rating of earthquake effects based, in part, on the energy they imparted to his devices as well as their actual effects on Louisville's well-being. He developed a six-part rating system ranging from "most tremendous" — a shock that threatened the destruction of the town — to one causing a "strange sort of sensation, absence, and sometimes giddiness," but otherwise not felt except by instruments. Later Robert Mallet used a coarser three-part scale — great, mean, and minor — to which he more or less arbitrarily assigned circular ranges. There were many other such scales devised here and there over the years. In Japan Fusakichi Omori (Milne's colleague) developed a scale that took account of structures peculiar to Japan, such as Buddhist temples. It was a seven-point scale that formed the basis of the official Japanese intensity scale used today.

The one in use in the United States today is based on a scale devised by an Italian seismologist and volcanologist, Giuseppe Mercalli, shortly after the turn of the twentieth century. It referred to Italian earthquakes and their effects on the Italian environment and was given in roman numerals from I to X. In 1931, it was revised by two American scientists, Harry Wood and

Frank Neumann, to take into account such things as progress in anti-earthquake construction. Called the Modified Mercalli Intensity (MMI) Scale, it continues to be revised, as the man-made environment changes. It specifies twelve levels of earthquake intensity with roman numerals. It is a finite scale: no earthquake can get any worse than a XII.

The MMI Scale is as follows:

I. Not felt except by a very few under especially favorable circumstances.

II. Felt only by a few persons at rest, especially those on the upper floors of buildings. Delicately suspended objects may swing.

III. Felt quite noticeably by persons indoors, especially on upper floors of buildings. Many people do not recognize it as an earthquake. Standing automobiles may rock slightly. Vibration similar to that of a passing truck.

IV. Felt indoors by many, outdoors by few during daytime. At night, some awakened. Dishes, windows, doors disturbed; walls make cracking sound. Sensation like a heavy truck striking building. Standing automobiles rocked noticeably.

V. Felt by nearly everyone; many awakened. Some dishes, windows broken. Unstable objects overturned. Pendulum clocks may stop.

VI. Felt by all, many frightened. Some heavy furniture moved; a few instances of fallen plaster. Damage slight.

VII. Damage negligible in buildings of good design and construction; slight to moderate in ordinary structures; considerable damage in poorly built or badly designed structures; some chimneys broken.

VIII. Damage slight in specially designed structures; well-designed frame structures thrown out of plumb. Damage great in poorly built structures. Fall of chimneys, factory stacks, columns, monuments, walls. Heavy furniture overturned.

IX. Damage considerable in specially designed structures; well-designed frame structures thrown out of plumb. Damage great in substantial buildings, with partial collapse. Buildings shifted off foundations.

X. Some well-built wooden structures destroyed; most masonry and frame structures destroyed with foundations. Railroad rails bent.

XI. Few if any masonry structures remain standing. Bridges destroyed. Rails bent greatly.

XII. Damage total. Lines of sight and level are distorted. Objects thrown into the air.

The reason the Mercalli scale is not used to describe earthquakes on television or in the newspapers is that by the time the quake has been rated, it is yesterday's news and no longer a pressing matter (except of course to those who lived through it and are still picking up the pieces). To assess a U.S. quake's place on the MMI Scale, the United States Geological Survey mails questionnaires to the postmasters in the disturbed area, requesting information. These and other data compiled by other means including investigations on the scene by engineers, geologists, and others are all put together and turned into, among other things, maps with isoseismal lines showing the extent of various levels of intensity. Such information is of vital importance to safety engineers and emergency management officials in planning proper development and construction in the area and also to students of earthquakes and their history.

In the early 1970s, a young geophysicist at St. Louis University, Otto W. Nuttli, undertook the task of mapping the intensity of the first New Madrid quake, publishing his results in 1973 in the *Bulletin of the Seismological Society of America*.[1] Nuttli wrote that with respect to the widespread area of perceptibility and damage, the three quakes "rank as the largest to have occurred in North America since its settlement by Europeans" and they are "somewhat of an enigma to those who are concerned with assessing or estimating the earthquake hazard problem in the United States." Nuttli and his associates found that east of the Rocky Mountains the most damaging waves during an earthquake attenuate — which is to say, lose intensity — at a rate that is about ten times *less* than those west of the Rockies. In part, this is because of the nature of the Mississippi drainage, a place of sandy soils that are largely waterlogged. But why such waves should attenuate so much less throughout the entire east than the west, he could not explain.

To map the intensity of the quake of December, he consulted both Myron Fuller's 1906 study of the New Madrid quakes and Samuel Mitchill's compilation of accounts. More thorough than even that, however, Nuttli went back to all of the available newspaper and other contemporary accounts, and assigned each account from each place a Mercalli Intensity unit. He researched accounts that appeared up until June 1812, explaining that while aftershocks persisted at least through the year 1813, they were "apparently not considered newsworthy, especially when compared with such events as the War of 1812 and Napoleon's adventures in Europe." Admitting the subjectivity of such a process, he asserted that the "uncertainty probably does not exceed ±1 intensity unit."

The map covers only the eastern range of the earthquake's effects since that is where the observers were. To the west, only Indians and maybe a few French or Spanish adventurers knew of the quake, but Nuttli in his commentary suggested that there was no reason not to extend the isoseismal lines to the west as well and with a similar radius. Thus, he estimated the area of poten-

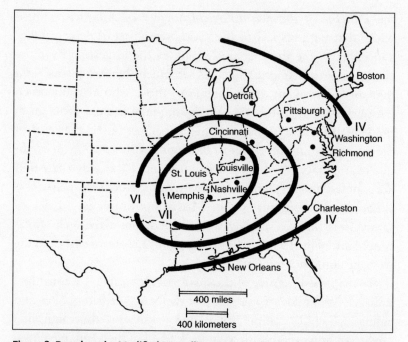

Figure 8. Based on the Modified Mercalli Intensity Scale, the concentric rings show the area of the most severe ground shaking and damage (VII), the area where the earthquake was felt by everybody but damage was slight (VI), and the area where the quake was felt by many but no damage occurred (IV). This intensity map was based on first-person accounts of the December 16 New Madrid quake, with the contours west of the Mississippi River extrapolated from observations to the east. Otto Nuttli, "The Mississippi Valley Earthquakes of 1811 and 1812," *Seismological Society of America Bulletin*, 63, pp. 227–248, 1973.

tial damage rated at VII as some 230,000 square miles. That is, if the entire region had been filled with poorly designed and badly built houses, the damage to all of them would have been severe and hardly a chimney would have been left standing. By contrast, the same intensity from the 1906 San Francisco quake extended over a mere 11,000 square miles.

Looking at another intensity unit — IV, where everyone would feel it, dishes and windows would break, and unstable objects would be overturned, presumably giving people something of a scare — the area reached by the New Madrid quake of 1811 was a *million* square miles. By contrast, the 1906 quake woke up people in an area of some sixty *thousand* square miles.

Having accomplished his map, Nuttli also wanted to establish the relationship between the intensity of the December quake and the other two immense shocks and the ground motion they gave rise to. This theoretically could be accomplished by comparing intensity at various distances from the epicenter with other quakes for which seismological data were available. But he couldn't use the data from the most carefully documented area in the United States, California, because western data were simply inapplicable to the East. But if he could establish the nature of the ground motion from the quakes he would then be able to assign the quakes a value on the Richter scale — that is, their magnitude. In fact, he worked this all out by various stratagems, and came up with the following magnitudes:

December 16, 1911	7.2
January 23, 1812	7.1
February 7, 1812	7.4

In other words, Nuttli wrote, "the combination of poor soil conditions in the epicentral area and of low attenuation of surface-wave energy produced damage and felt areas 100 times greater than those of western North America earthquakes of the same magnitude." He warned of complacency in the midsection of the country. In fact, and as we will see, Nuttli's magnitude estimates were far too low. (He would later make new estimates that turned out to be far too high.)

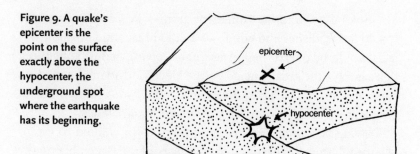

Figure 9. A quake's epicenter is the point on the surface exactly above the hypocenter, the underground spot where the earthquake has its beginning.

Since the MMI provides valuable information, why, one might ask, is the Richter scale needed? What more does anyone really need to know about a quake besides the extent of the damage, that is, its *intensity* over a given area, which is something, as we have seen, the Richter scale does not tell you? In fact, the Richter scale does not measure any particular physical quantity, such as energy or temperature or acceleration, or even a quake's actual *power*. Its numbers do not refer to such units as joules or kilowatt-hours, by which amounts of work done are measured, though of course an earthquake does a great deal of what a physicist would call work. The Richter scale measures earthquakes in "magnitude units," which are basically artificial constructs. Indeed they are artifacts produced by a set of mathematical rules having to do with the maximum deflection of the stylus (or other recording device) of a seismograph, adjusted for the distance of the seismograph from the exact place underground the quake began (the hypocenter, the epicenter being the exact place on the earth's surface above the place where the quake began).

The Richter scale came about when, in the late 1920s and early 1930s, Charles Richter was a young physicist working in the Seismological Laboratory of the California Technical Institute. He and the director (Harry Wood, who would later modify the MMI Scale) were working on a historical catalogue of Southern California earthquakes and also measuring seismic waves in the region with an array of seven widely spaced stations. It suddenly struck Richter that earthquakes could be compared in terms of the measured amplitudes recorded at the seven stations — corrected for distance. A Japanese seismologist, Kiyoo Wadati, had done something similar — relating ground motion to distance in the case of very large quakes — but when Richter tried to do the same, he found that the range between small and large quakes gave figures that were "unmanageably large." A colleague, Beno Gutenberg, who had recently relocated to Caltech from Germany, suggested that Richter use logarithms instead of plain numbers. What many today still are unaware of is that since the scale is logarithmic, a Magnitude 2 quake is *ten times* the "size" of a 1, a 3 ten times a 2, and so on. This is not, as we pointed out, a measure of the actual energy released by an earthquake: as it turns out, though, the energy released is greater by about *thirty to fifty* times each step up the scale.

And thus, in 1935, was born the Richter scale. It was originally set up only to compare Southern California earthquakes and only with the use of the particular seismographs in use in the Caltech array of seven. A Magnitude 0 quake is the smallest tremor Richter imagined we would ever be able to record. A 3 was the smallest earthquake one could feel; and 5 through 7 were the largest quakes that Southern California would be expected to ever suffer. These units were expressed as whole numbers and tenths.

The Richter scale came into wide use once it was adapted to work worldwide and with various kinds of seismographs. These complex mathematical adaptations were worked out by Richter's colleague Gutenberg, who was probably the world's leading observational seismologist. The Richter scale is different than an intensity scale in more ways than one. In the Mercalli scale XII is tops: nothing is worse. It is like the scale of 1 to 10 people use for all sorts of comparisons. In expressing one's opinion about, say, a movie, a 1 is as bad as it gets, and there is nothing better than a 10. The Richter scale, on the other hand, is open-ended at both ends. There are earthquakes that come in at 8s and even 9s, and the only upward limit is probably the limits inherent in the earth itself. Meanwhile there could, at least theoretically, be a −2 tremor: it would be one hundred times smaller than the smallest one Richter imagined ever being recorded.

For mostly mathematical reasons that are more complicated than nonseismologists need to know, the upper end of the Richter scale gave inconsistent measures, so it was modified to more accurately reflect the nature of the larger (7-plus) quakes. And in recent years it has essentially been retired by seismologists. When the press comes calling, demanding to know the magnitude of a quake on the Richter scale, the scientists will give a reasonable magnitude number, and the press may call it the Richter scale magnitude or not. More often than not today, it is simply given as the "magnitude." For their own purposes, however, seismologists have developed a scale that directly measures the actual characteristics of a quake. It is called the seismic moment.

In the meantime, to have pointed out that the Richter scale eventually proved unsatisfactory is by no means to denigrate the man's achievements as a seismologist. He did a great deal of valuable work in the field, much of it in collaboration with Beno Gutenberg, and Richter was evidently always a bit sheepish about having his name so firmly and singularly attached to a scientific achievement to which Gutenberg had contributed a major idea (the logarithms). Gutenberg was, in fact, what today might be called a scientific superstar. He was born in Germany in 1889 and by the 1930s was considered by many to be the leading seismologist in the world. His best-known contribution to the field of geophysics was the location of the earth's core — about 1,740 miles below the surface. To avoid the growing anti-Semitic sentiment in Germany he accepted a position with Caltech in the

United States, and went on to coauthor with Richter four monumental treatises on seismic waves, covering such things as the travel times for waves, deep-focus earthquakes, amplitudes, surface waves, microseisms — in all, the bible for seismologists. He discovered very long-period seismic waves that circle the earth, as well as the differences in the crustal structure between the oceans and the continents. He was what could be called the Compleat Geophysicist. When he died in 1960, one of his American colleagues said of his prolific career that "it is rare that anyone writes a paper in seismology without referring to him."

Seismic Moment

Beno Gutenberg and others had worked out the relationship of magnitude to energy — the amount of energy was calculated to rise thirty to fifty times with each upward numerical step of magnitude on the Richter scale.[2] The energy involved in an earthquake of Richter 5.5, for example, was calculated to be about as much as that released in the atomic bomb blast on Bikini in 1946.

But the energy measurements calculated to correspond to the Richter scale were based on the maximum single ground-wave amplitude as recorded on a seismograph — just as was the scale itself. Seismologists began to realize that this was something like judging the power of a storm by measuring its strongest wind gust. Something more precisely anchored in the actual physical events of an earthquake was needed — something that would provide a consistent absolute scale of earthquake size.

Suppose, as some teachers suggest in explaining this sort of thing, that you put both hands on the edge of a table. Then, with your right hand you push the table away from you while, with the left, you pull it toward you. The table rotates. If your hands are close together, it takes a lot more force to rotate the table than if they are farther apart. The force you need to apply to the table to rotate it is called torque, a term familiar to auto buffs and others. It is a mechanical phenomenon, and so is an earthquake. The torque of an earthquake is called the seismic moment, and it refers directly to an earthquake fault and the amount of force that causes it suddenly to slip, leading to its elastic rebound all along it and across it, not unlike what happens when a stretched rubber band is released. It is the amount of energy set off by the rebound that is meant by the term "seismic moment," which is sometimes simply referred to as "moment."

The way it works out, this involves a great deal of complicated mathematics too, but it essentially is reached by multiplying three values: the area of the rupture (which could be hundreds of square miles), the slip produced by the earthquake (the distance of the displacement), and something called shear resistance (which is a measure of the material strength or rigidity of the rock involved). Giving earthquake size in these terms has turned out to be an accurate and absolute way of measuring earthquakes and comparing them, but it is not the sort of thing that a television newscaster is likely to cotton to because it gets a bit unwieldy. If an earthquake so small that it can only barely be felt at all is given a value of 1, the largest earthquakes that ever happen might be something like 1,000,000,000. So once again, a logarithmic scale is applied, bringing the numbers down to something a bit more manageable.

In this way, the seismic moment conforms relatively closely to the Richter scale in magnitudes ranging from Richter 2 through 5. At higher magnitudes, the seismic moment (adjusted with logarithms) is a more faithful reflection of an earthquake's size than Richter. Most reports of earthquake magnitudes are now based on the seismic moment and are referred to simply as "magnitude," without anyone's name attached. In scientific journals the number is preceded by **M**. (Unless otherwise noted, earthquake magnitudes provided henceforth in this book are all **M**.)

As a result, a number of well-studied and well-known historical earthquakes have had to be scaled back — or forward. In the 1980s, the San Francisco quake of 1906 was considered Richter magnitude 8.2. It is now taken to be 7.7. The three humongous New Madrid quakes are classed as low 8s but for one of the three that is a high 7, though some disagree and think of them as all in the upper 7 range. The difference (in power) of a low 8 and a high 7 is considerable; on the other hand, the difference between a low 4 and a high 3 is comparatively trivial. (Making magnitude estimates, especially for big quakes, has extremely important practical implications, as we will see in due course.)

In the bustle of actual practice, calls for immediate magnitude ratings also can lead to a high number that is revised slightly downward once the dust has settled (as it were), and this can be confusing. As Susan Hough has written, it can convey "a less-than-favorable impression of the collective competence of seismologists." The layman needs to remember, however, that every earthquake, like every snowflake, is different from all the rest. And each one is something like an instantaneous game of three-dimensional chess played in a distant room.

7

Geophysical Leaps Forward

AMONG THE MANY SCIENTISTS who turned their attentions to the needs of a nation waging World War II, Beno Gutenberg put his talents in the service of the United States Navy. Among other accomplishments, he had determined that microseisms were not so much the product of surf as of storms raging through the ocean depths. The result was the tiny seismic disturbances that are a main cause of the constant background buzz that seismologists have to cope with. From this, Gutenberg could help forecast storms, especially hurricanes, and in particular the hurricanes that arise in the Caribbean and that have plagued navies and other maritime craft since the Spanish arrivals after 1492.

Seismology was by no means the first or only science to enter into high gear as a result of American military needs. The U.S. Navy funded all sorts of oceanographic research in the war and afterward, much of it to benefit the submarine corps. Much of this involved mapping the sea bottoms, but it also funded research into the vocalizations and other sounds created by sea creatures such as whales, the better to operate the submariners' crucial listening system, sonar. Much of the oceanographic work would wind up being of direct benefit to the earth sciences and in particular to earthquake science, but nothing would benefit the practice of seismology in the latter half of the twentieth century as much as the military's need to monitor the various test ban treaties that sought to tame the nuclear arms race.

A decade after the first atomic bomb was exploded at a site called Trinity in the arid desert lands of New Mexico on July 16, 1945, public pressure to put an end to testing nuclear weapons led to the Limited Nuclear Test Ban Treaty in 1963. A comprehensive test ban treaty had to wait because seismological capability at the time was deemed inadequate for monitoring underground tests. To begin with, the science of seismology was not widespread, and only relatively limited global information existed. Seismic stations were relatively few and far between and operated at a limited range of frequencies.

One effort had already been made to see if nuclear test explosions could be contained underground, with the ancillary function of observing its seismological effects. This test was dubbed Rainier, a detonation of a 1.7-kiloton nuclear device at the far end of a spiraling tunnel in Rainier Mesa, Nevada, on September 19, 1957. The Rainier test produced strong seismic signals that, at the time, appeared much like those of an earthquake, and from that moment on, the science of seismology has been directly enmeshed in the crucial business of nuclear nonproliferation.

The nuclear nations were nearly paranoid about each other, especially the United States and the USSR. Knowing what each other was doing in this realm was of the most crucial importance, and with testing being done underground, what was needed was a complete modernization and upgrading of seismographic capacity around the world. This effort came to be known as the

U.S. government's VELA program. It was funded in large part by the Advanced Research Projects Agency of the United States Department of Defense.

By the early 1960s, the first World Wide Web had been established. Called the World-Wide Standard Seismological Network, it consisted of 125 stations in thirty-one nations around the world. Of the utmost importance was that these seismographs were not just state-of-the-art but all compatible. An equally important effort was a major research effort to understand in ever greater detail the generation of seismic waves from nuclear explosions and other seismic events and how they propagate through the earth in various regions. Over the next three or four decades, thousands of papers were published in peer-reviewed journals by university and government seismologists — research that extended the capabilities of seismology by what can be thought of as at least an order of magnitude.

The seismographic stations deployed in the 1960s for this purpose recorded data on magnetic tape that had to be collected periodically. By the late seventies, Sandia National Laboratories had produced stations that could transmit ground motion data via satellite. Today, the seismographs and the transmission systems are all electronic, open, and tamper-proof. New seismographs operate on broadband, making it possible for the same instrument to record and transmit both low- and high-frequency waves. So fast is this system now that data from a seismic event may arrive at far-distant stations before the P waves themselves.

Back in 1974, the Threshold Test Ban Treaty was signed, restricting the yield of nuclear tests by the United States and the USSR to less than 150 kilotons. Bugsy Siegel, the founder of Las Vegas as a gambling and mob haven, would have been pleased. The yield of 150 kilotons was chosen with a view to the level of shaking caused in that city by testing at the Nevada Test Site. In any event, seismology was now involved not merely in detecting underground nuclear tests but also in determining their yield (or magnitude) — a different kind of challenge.

Some parts of the job of distinguishing underground nuclear tests from earthquakes are not that difficult. First of all, some seismic events can be pegged as earthquakes simply by the fact that the hypocenter (the place underground where the quake originates) is too deep. No one has ever drilled a hole deeper than about six miles, and most underground nuclear tests occur at around a mile down or less. Most earthquakes start much deeper than six miles. As noted earlier, earthquakes take place usually when rocks slide against each other over a wide area, sending out a variety of seismic waves in various directions over measurable numbers of seconds — a far cry from a nuclear explosion that releases its energy in less than a millionth of a second.

The best way to hide such a test, it was discovered, was to suspend the device in a large underground cavity. This "decouples" the blast from the surrounding rock. The stresses on the rock do not exceed the elastic limit of the rock, which means that any seismic waves are generated into the surrounding earth inefficiently. The evident effects are to reduce the apparent size of the blast by a factor as great as seventy, but often less than that. To damp the effects of a five-kiloton explosion, one needs to excavate a spherical hole deep in the ground with a diameter longer than the Statue of Liberty is high. Even with explosions damped in this manner, however, making them appear industrial in nature, seismologists had learned by the 1990s to determine if such carefully camouflaged explosions were nuclear or not.

Deploying all this military-backed seismological science in behalf of peace has helped make the field more exact than John Milne and all the other early practitioners could have dreamed. The benefits to understanding earthquakes (and volcanoes) and to learning about the internal structure of the planet have been immeasurable. It was data collected by the World-Wide Standard

Seismological Network back in the sixties that helped bring about the greatest single revolution in the science of geophysics — and led to a nearly complete understanding of what makes earthquakes happen at all. The phenomenon was first called continental drift.

Continental Drift

Many books on the subject of the earth sciences — both popular and technical — that were published in the late 1950s and early 1960s mentioned the notion of continental drift, and most of them referred to it as a bit harebrained. Just what could cause these huge slabs of rock to drift around the surface of the planet was inexplicable, and without a mechanism, a driving force, the idea remained fanciful at best. But almost overnight, practically every earth scientist did a one-eighty. By the late 1960s, the idea of continental drift — redubbed plate tectonics — was agreed upon by geophysicists as the explanation of the basic nature of the planet's history, and also the underlying reason for most earthquakes. There were, to be sure, a vanishing few holdouts in the geological community but they soon disappeared from view altogether, leaving only a waggish group known as the Stop Continental Drift Society which issued a whimsical newsletter replete with schemes for screwing down the continents so that they would stay put, and adorning a few cars with bumper stickers that pleaded, "Stop Continental Drift." In all, the acceptance of plate tectonics was one of the quickest revolutions in the history of science. On the other hand, it surely did not seem that way when a German meteorologist forcefully presented it to the scientific community.

He wrote that the continents of the world had once formed a single great landmass and had begun drifting toward their present configuration some two hundred million years ago. He wrote at the time when the War to End All Wars, also known as the Great War and later as World War I, was still raging across Europe. The author, Alfred Wegener, had earlier been sent off to the front, where he was promptly shot in the neck and, while recovering from this wound, wrote *The Origin of Continents and Oceans*. For the most part, the book was ignored in Germany and, when translated into English and other languages, derided as "mere geopoetry." As science, it was seen as inept, and of course it challenged the very fundament of the geological thinking at the time — or at least most geological thinking — and worse, it was the proposal of a man trained as an astronomer and a meteorologist — not a proper geophysicist or geologist.

Even so, in 1926, a symposium on the subject was held at the annual meeting of the American Association of Petroleum Geologists, and Alfred Wegener was invited but did not attend, and probably for good reason. Petroleum geologist after petroleum geologist took the podium and ridiculed him and his theory. A University of Chicago scientist questioned if geology could still be regarded as a science if it were "possible for such a theory as this to run wild." Another attendee, from Johns Hopkins, arose to say that Wegener's methodology "is not scientific but takes the familiar course of an initial idea, a selective search through the literature for corroborative evidence, ignoring most of the facts that are opposed to the idea, and ending in a state of autointoxication in which the subjective idea comes to be considered objective fact."

We now know that Wegener was far more than just a meteorologist/astronomer. He was a remarkably energetic, innovative, and adventuresome man who was also deeply self-trained in other branches of the earth sciences including paleontology. And of course he turned out to be right. Alfred Wegener is properly considered to be the father of the theory of continental drift, having made the first sustained effort to persuade others it was true (though like virtually all robust theories it has multiple sires too). His main problem in the 1920s was that neither he nor anyone else at the time could imagine what yet unknown force could possibly cause something the size of continents to move at all, much less "plow through the mantle," as someone put it. But Wegener was confident that he was right, and knew for a certainty that the movement of the continents would one day explain a great deal. "It is probable," he wrote in a 1929 revision of his book, "the complete solution of the problem of the forces will be a long time coming. The Newton of drift theory has not yet appeared." But he went on to assert with stunning prescience that the "forces which displace continents are the same as those which produce great fold-mountain ranges. Continental drift, faults and compressions, earthquakes, volcanicity, [ocean] transgression cycles and [apparent] polar wandering are undoubtedly connected on a grand scale."

The idea of continental drift had not sprung fully formed from Wegener's brain as Athena sprang spontaneously from Zeus's forehead already armed, icy, and wise. Its antecedents — what might in the name of geopoetry be thought of as barely perceived foreshocks — began to occur much earlier. Not surprisingly, the first proposal of continental drift was made in the late 1500s by a cartographer, for who else but a mapmaker could have perceived the wonderful fit of the continents' shapes any earlier? The cartographer was Abraham Oertel of Antwerp, a self-described Belgian-German.[1] Oertel became a cartographer honestly: his father died when he was a boy, and he helped the family survive by what might be thought of as carto-hustling. He bought charts and maps, mounted them on canvas, colored them, and resold them. He became a full-fledged mapmaker in his twenties, soon publishing an eight-leaved map of the world five years before Mercator published his world map. No political naif, Oertel, now calling himself Oertelius, dedicated this first great modern atlas, *Theatrum orbis terrarium,* published in 1575, to the king of Spain, who, deeply moved, made him the royal cartographer, from which position Oertelius had access to a great deal of geographical information gathered by Spanish explorers and which the Spanish Crown considered as proprietary. In any event, Oertelius had already noted the similarity in shape of the eastern and western margins of the Atlantic Ocean and suggested that the Americas had been "torn away from Europe and Africa . . . by earthquakes and floods."

As maps of the increasingly known world became more common, it is almost certain that others came to the same conclusion, but the next one whom we know of was none other than Sir Francis Bacon, the juridical scholar, lawyer, and philosopher who famously pointed out that science should be a matter of reasoning from careful observations of the natural world, rather than spinning pretty theories without benefit of such observations. In 1620, in something of a throwaway observation, he noted the matching shapes of the continents but took the thought no further. Ben Franklin later suggested that the earth had a fluid core that buoyed up the continents on its rocky shell, and the great French naturalist Georges de Buffon and the German geographer Humboldt, whom we met earlier, also made the same sort of suggestion.[2]

Then, in 1858, in a book modestly titled *Creation and Its Mysteries Revealed,* a French geographer, Antonio Snider-Pellegrini, said that the Americas and Africa had at one time been joined and were torn apart during Noah's Flood and the stresses it must have caused, producing among other effects an asymmetric distribution of mass in the earth. This suggestion was in one way more scientific than anything like it earlier, for Pellegrini suggested that the hypothesis could be tested: one merely had to determine if the coal-rich rocks on either side of the Atlantic would indeed line up perfectly if the continents were "put back" in their original, joined configuration. On the other hand, he announced all this well after Charles Lyell's geology had pretty much discredited Noah's Flood and other catastrophes as the driving forces in earth history. Perhaps this was sufficient reason for the world to pay no attention to his theory — making it what we might think of as yet another largely imperceptible wiggle on the seismogram of geological history.

More noticeable was the journey of HMS *Challenger,* the first great oceanographic expedition, a joint project of the British Admiralty and the Royal Society that spanned four years beginning in 1872 and traversed almost seventy thousand nautical miles. Among its measurements of ocean temperatures, depths, and currents throughout the world, it produced the first survey of the ocean basins that account for more than two-thirds of the planet's surface. Along the way they noted that huge submarine ridges were to be found in the oceans, including one that runs north-south in the middle of the Atlantic Ocean.

It would be nearly a century before anyone had much of an idea about what this Mid-Atlantic Ridge was all about, or the others like it elsewhere in the deep, or to connect them to the idea of continental movements. But, meanwhile, fossil ferns were causing some interesting speculation. In 1875, an Austrian geologist, Edward Seuss, began publishing a multivolume series called *The Face of the Earth* in which he took note of the nearly worldwide presence of fossils of the fern genus *Glossopteris.* Europeans had long found similar species of this type of fern in European and North American sites, but with the expansion of European colonists into other parts of the world, an entirely different *Glossopteris* flora was turned up in India, Australia, South America, and even Antarctica. Noting that ferns, like most plants, are relatively sessile and not given to transoceanic migrations, Seuss took these two different floras as a sign that the southern continents and India had once been connected by what he posited were extensive land bridges. The continents in his view had not moved: instead, at some time in the deep past, the land bridges had all sunk into the oceans.

Just why the land bridges would have all sunk could be answered with some logic, given the times. One of the great physicists of all time, Lord Kelvin, had recently suggested that the earth was slowly cooling off from its fiery, molten origins, and as it cooled it of course had to contract. This shrinkage would have caused lateral compressive forces across the surface of the earth, Seuss reasoned, that not only would have sunk the land bridges but forced parts of the surface up into folded belts of mountains like the Alps and the Himalayas.

Shortly after the turn of the twentieth century, Frank Taylor, a glacial geologist with the United States Geological Survey, wrote a paper that was largely ignored at the time in which he argued that the contraction of the earth could not account for the huge lateral displacements of rocky material to be seen in the major mountain belts of the world. Instead, he proposed that Europe and Asia were sliding southward and the Americas sliding to the west. The friction and compression of such movements (which he specified were very, very slow) would have caused the leading edges of these continents to rise up as mountain ranges. Not only did the similarity of fossils across oceans suggest this but so did geological formations and faults that also matched. More or less contemporaneously, another Austrian, Otto Ampferer, also found the contraction hypothesis inadequate, and suggested that thermal convection in the earth's interior caused warm mantle material to well up under the ocean basins, while cool mantle material sank under the continents. The net effect would be that the sinking mantle material would pull continental material together, making it bunch up into mountain belts. Like Taylor's, Ampferer's speculations went ignored.

Yet, for an earnest and dogged researcher (and maverick) like

Alfred Wegener, enough strange and wonderful transoceanic similarities had been brought to light over the years to let him take the unpopular view that the continents had to have drifted apart, regardless of the fact that there was no reasonable explanation of what could have caused such motion. On the other hand, there were reasons as well to dismiss explanations of the *Glossopteris* phenomenon and others — particularly the land bridge idea of Herr Doktor Seuss.

There is a misperception generally about what is meant by a land bridge — like that across the Bering Strait during the Ice Age when the sea level fell, exposing dry land from Russia to Alaska. This was not some narrow isthmus across which mammoths and early human colonists picked their way in earshot of crashing surf on both sides, but a chunk of land a thousand miles across from north to south. Presumably, the land bridges Seuss posited were something similar. And Wegener, though not trained specifically as a geophysicist, knew enough in that realm to know that Seuss's land bridges simply could not have been pushed downward by some force or another and stayed put. According to a theory called glacial rebound, the land bridges, if shoved downward, would have risen again.

They knew about this in Wegener's time because it had been discovered that the lands of Scandinavia had gradually been pushed downward some 250 yards by the accumulation of the glacier's weight as it grew to its maximum size (almost two miles thick in places). As the ice began to melt, which it did relatively quickly, retreating to about its present northern position some eleven thousand years ago, the once glaciated lands began to slowly rise, and have achieved about two-thirds of their original elevation by now. The process is still going on, in other words. What had happened in this down-and-up motion was that the hot rock of the mantle had spread slowly out and away from the center of the increasing weight of the glacial ice and, as the ice disappeared, it had begun to flow back in under the original land, pushing it back up. The slowness of the process meant the underlying warm rock of the mantle was indeed viscous. Had the rock underlying the glacier been perfectly elastic (i.e., not viscous), the land would have been slightly depressed under the load of the ice and then, with ice gone, it would have bounced rapidly back (something in the manner of an earthquake's suddenness). In either case Seuss's land bridges simply could not have been shoved permanently down to the ocean bottom.

Also, by Wegener's time it had been well established that in an ancient period known as the Carboniferous, glaciers had also come and then gone over the lands of South America, Africa, Australia, India, and Antarctica. This made no sense whatsoever if all these bodies of land had been in their present configuration on the surface of the planet — the glacial ice of the Carboniferous would have occurred in a haphazard and unexplainable patchwork. On the other hand, if the southern continents of today had all been one landmass, the glacial regions and other ancient climatic zones would have all lined up perfectly. Some two hundred million years ago, Wegener suggested, this large southern continent called Gondwanaland (from a term that Edward Seuss derived from an Indian dialect) split apart and began drifting into the present configuration of the pieces. And before that, Wegener hypothesized, Gondwanaland had been part of a yet larger landmass he called Pangea (from the Greek words meaning, approximately, all the land). Pangea, then, had separated into two supercontinents — the aforementioned Gondwanaland of the south and Laurasia of the north, the latter being today's Europe, Asia, and North America.

The idea of the continents drifting around willy-nilly seemed ludicrous to most professional geologists. Like most other fields of science at the time, geology had become highly professionalized and replete with specialties and subspecialties. It was, in other words, a mature science, a corollary of which is that the opinions of people not specifically trained in its specialties are usually not welcome. So Wegener was ridiculed by the American geologists and his theory rancorously denounced by most of the geological community. Indeed, the very virulence with which it was damned may have been a gut reaction to some of the undeniable but embarrassing phenomena it raised — phenomena that could not be aptly explained by the then accepted tenets of geology and therefore had to be studiously ignored. A few attempts to explain these embarrassing phenomena did ensue — mostly calling forth events and other phenomena that were equally or more implausible by contemporaneous standards, and for the most part, the matter was simply dropped. The geological world was ruled by what are called the Fixists — people with the adamantine belief that the continents were and always had been fixed in their present locations. To be fair, what we today call defining data were missing from the notion of continental drift — the force great enough to move continents and the mechanism by which such a force was applied.

Wegener died before his time. On a scientific mission to Greenland in 1930 (where he had earlier in his career done some pioneering meteorological work with balloons) he was attempting to bring aid to some stranded members of his party and froze to death at the age of fifty. Had he lived into his eighties, he would have enjoyed his vindication, for even as he perished on Greenland, a crescendo of new geophysical discoveries began that would culminate in a grand rush of insights and the acceptance of an entirely new geological paradigm.

In the late 1930s, a Japanese geophysicist, Kiyoo Wadati, had seen that most of the world's earthquakes occurred on land in the major mountain belts like the Alps, Himalayas, and Andes, and along submarine mountain ridges that had been perceived, if only dimly, as early as the 1870s. Wadati's studies were furthered by Hugo Benioff of the California Institute of Technology in the years after World War II. With the benefit of improved seismographic instruments, some of his own invention, he studied the quakes that occur often along the Kamchatka Peninsula and northern Japan. Just offshore is a deep ocean trench and onshore a line of volcanoes. Just below the bottom of the trench, Benioff discovered that faulting produced major quakes, which occurred at increasing depths as one moved (via seismographic readings) westward into the interior of the Asian continental shelf. No one at the time could explain this phenomenon, which was soon found to be common around much of the edge of the Pacific. Such places are called Wadati-Benioff zones.

Meanwhile the greatly admired geophysical polymath Beno Gutenberg had taken an active interest in the topic of convection in the earth's mantle. Put simply, he and some others had concluded by the early 1950s that because of differing gravity effects on different parts of the earth's mantle, it flowed up in some places and down in others. Gutenberg believed that such motions could well be a potent enough mechanism to power continental drift. But even Gutenberg's interest in the theory was insufficient to move the great proportion of geologists into the drift camp, which was a small group sometimes called Mobilists.

In a different realm, for several decades the earth had been understood to act as a gigantic magnet — thanks at least in part to the rotation of its metallic core. The lines of force it produces — its magnetic field — run vertically out from the South Pole in extremely wide arcs until they are pulled vertically back to earth at the North Pole. Now, it was known that when some molten rock cools, little particles of a mineral called magnetite orient themselves along the lines of the magnetic field and are fixed in position permanently. If the rock had cooled at exactly the North

Pole, the little magnetite particles would be oriented directly up and down. If the cooling rock was at the equator, the particles would be lined up exactly parallel with the earth's surface, since at the equator the magnetic force lines are parallel with the earth's surface. Indeed, one should be able to tell what latitude a piece of rock came from by seeing what angle its magnetite was, since it would mimic the angle of the magnetic field at that latitude. But in the 1950s, scientists found that, on every continent, the older the rock analyzed, the more its magnetite's angle differed from what was the present angle. In other words, it looked like the continents had to have moved.

Meanwhile it had become clear that the ocean bottoms were characterized by stripes of magnetic polarity that showed that every so often the earth's magnetic field quite abruptly does a one-eighty. It changes from south-to-north to north-to-south — and for reasons that remain unclear. The intervals between these still mysterious switches can be tens to hundreds of thousands of years, and a time chart was drawn up for the past few million years showing periods of normal and reversed magnetic direction. The postwar U.S. military had not just summoned up better seismographs, but also more sensitive magnetometers, the better to look for submarines. And these instruments would play another role altogether.

Not long after the end of the war, Maurice Ewing and Bruce Heezen, both based at Columbia University, produced a map of the Indian, Atlantic, and Antarctic oceans showing a mostly continuous valley along the mid-ocean ridges, what they took to be the source of the seismicity in these mid-ocean areas that had been documented earlier by Gutenberg and Richter. Then, in 1963, Fred Vine and Drum Matthews, both of Cambridge University, published information in the British magazine *Nature* that sought to explain the known pattern of magnetic "stripes" that paralleled the great ridges on the mid-Atlantic floor. Each stripe represented an era of magnetic polarity opposite from the adjacent stripes. This, they said, could be explained only by the sea floor itself spreading out from the ridges. Thus the spreading of the sea floor could be shown in a known chronological sequence. (A similar explanation by Lawrence Morley, a member of the Canadian Geological Survey, had been submitted to two scientific journals earlier than Vine and Matthews's paper was sent to *Nature,* but had been rejected.) In any event, here was a mechanism, at least dimly perceived, that could explain what drove the continents to drift across the surface of the earth.

Suddenly, not unlike an earthquake, in the next four years, it all came together. Robert S. Dietz, a marine geologist at the Scripps Institution of Oceanography in California, had published a paper in *Nature* in 1961 that said the mechanism was, as Ampferer, Gutenberg, and others had said earlier, a matter of convection. Hot material below the surface (being less dense) rises at the mid-ocean ridges. It pushes the ocean floor away, and as it cools, it sinks somewhat. Volcanoes would occasionally form at the ridges and in some instances stick up above sea level to be worn down by erosion and at the same time sink somewhat along with the cooling sea floor beneath them. This explained the inactive flat-topped seamounts (called guyots) that had been discovered here and there in the Pacific during World War II. At the same time that the sea floor was spreading at the ridges, the Wadati-Beniof zones with their deep earthquakes around the margin of the Pacific Ocean could be seen as places where the mantle's convection cells were downwelling, causing the old ocean crust to dive back down into the mantle. (There had been some argumentation that if the earth's crust was steadily being formed at the mid-ocean ridges, it would mean that the earth was getting bigger — which of course it isn't.)

A Catastrophic Proof

At 5:36 P.M. on Good Friday, March 27, 1964, the second largest earthquake ever recorded by instruments occurred some twenty miles beneath northern Prince William Sound in Alaska and rang the entire earth like a bell. This was the famous Anchorage quake. It traveled southwest down the Alaska coastline and along the Aleutian chain and its associated deep trench. It hit Kodiak Island four hundred miles away in about four minutes. The huge shaking caused an enormous "tidal wave," more properly known as a tsunami, since they have nothing to do with the tides. At the Valdez waterfront a hundred yards of dock disappeared as the tsunami rose up about forty feet, followed shortly by another. No one in the waterfront area survived. At Kodiak Island the tsunami reared up more than sixty feet and destroyed much of the downtown as well as half the fishing fleet. The tsunami rampaged as far south as Crescent City, California, near the Oregon border. In the city of Anchorage, seventy miles west of the epicenter of the quake, one observer wrote that "table-top pieces of earth moved upwards, standing like toadstools with great overhangs. . . . A chasm opened beneath me. . . . I tumbled down. . . . Then my neighbor's house collapsed into the chasm." Some two thousand landslides and avalanches were attributed to the quake, along with 130 deaths and some $300 million in property damage.

Investigators, observing the displacement of beaches and seeing stranded colonies of sea creatures like barnacles, mapped a pattern of unusually large vertical motions along the coastlines — uplifts reached more than thirty feet, with depressions here and there of six feet. Later, researchers estimated that some 120,000 square miles of the Pacific Ocean crust had been thrust upward, the greatest area of vertical displacement ever recorded. Matching these patterns with seismological data from elsewhere including the San Andreas Fault, they found themselves with an uncannily timely textbook example of what geologists were just beginning to see with some clarity: plate tectonics. For what had happened was the instantaneous rotation of the Pacific and North American plates. In the process, the edge of the Pacific plate suddenly heaved upward and then dove downward, thrusting itself under the less mobile crust of the North American plate with a tremendous outburst of energy. The Good Friday quake was given a moment magnitude of 9.1, making it a hundred times more powerful than the San Francisco quake of 1906. In the twentieth century, only a quake in Chile in 1960 was more powerful (seismic moment magnitude 9.9). The main findings on the Good Friday quake were published by 1967.

By 1965, Tuzo Wilson had already shown that the surface of the earth consisted of several more or less rigid plates of crust that move with respect to one another. Where they abut one another and move in different directions are what are called transform faults — which was the case on Good Friday in Prince William Sound. Continents themselves are carried along on these plates and, being less dense and more buoyant, do not get hauled back into the mantle. Where two continents collide head-on, as was the case when the present subcontinent of India split off from Gondwanaland and crashed into the Eurasian plate, the earth shoves and buckles upward into vast mountain chains. The Himalayas are still being created all these forty or so million years later.

By 1967, a huge tectonic shift had suddenly taken place in the world of geology. ("Tectonic," by the way, literally means "pertaining to construction," and it derives from the Greek word for carpenter.) In just a matter of years, a wholly new version of the nature of the earth was constructed from what had been mere geopoetry.

Put simply, heat within the earth generated by the primal

hot and liquid core of the earth and by the decay of radioactive materials creates the convective forces of the viscous part of the earth's mantle, which in turn impels the great rigid slabs of the crust to move laterally. The upwelling of magma along the mid-ocean ridges (the spreading zones) causes about 5 percent of the release of the earth's seismic energy. The ridges are fed by molten material that rises up as giant mantle plumes arising presumably from around the boundary between the lower mantle and the liquid core — about halfway to the center of the planet. A frequent result is volcanoes, some of which rise up far enough to become islands.

To look a little more closely at this phenomenon, the rising molten material in these plumes is basic (as opposed to acidic), dominated by heavy metals and enriched in sulfur dioxide along with carbon dioxide, chlorine, and water. These volatile constituents of the magma are vaporized when the molten magma erupts at the surface.

It stands to reason that if two plates are moving away from each other in one place, they will be crashing against something else at the other, leading end, and this is what happens. The real action — 90 percent of the earth's seismic energy — occurs where the plates run into each other. Sometimes this is like a head-on collision as with the Himalayas and the Alps. Sometimes the collision is more of a slow-motion sideswipe, the aforementioned transform faults. The most common result is subduction, where one plate descends back into the mantle. When the edge of a plate is subducted to a sufficient depth, its material, enriched with water, reaches temperatures high enough to bring about at least partial melting, which in turn produces chambers of magma, which tend then to rise up. In other words, volcanoes, usually accompanied by deep-earth quakes.

While sea floor spreading appears to be relatively constant,

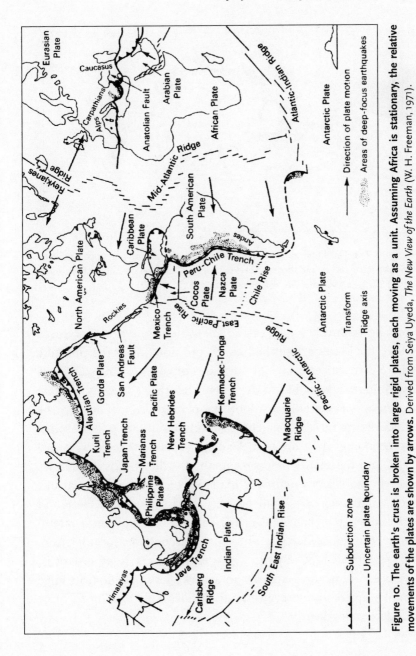

Figure 10. The earth's crust is broken into large rigid plates, each moving as a unit. Assuming Africa is stationary, the relative movements of the plates are shown by arrows. Derived from Seiya Uyeda, *The New View of the Earth* (W. H. Freeman, 1971).

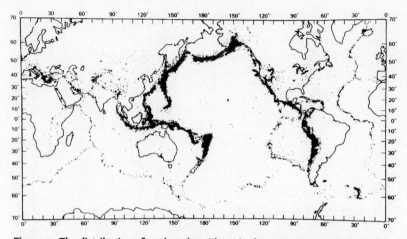

Figure 11. The distribution of earthquakes. Those in the oceans are concentrated on the mid-ocean ridges and those around the Pacific Ocean on the subduction zones. Derived from Seiya Uyeda, *The New View of the Earth* (W. H. Freeman, 1971).

creating relatively small earthquakes along the boundary area where the plates are pushed apart, the situation is the opposite where plates meet. Maps of the plates often show fairly uniform and smooth edges, and to the extent that they suggest that plate tectonics is essentially a smoothly operating conveyor belt, the drawings are misleading. There is nothing smooth where a plate is descending below another. An oceanic plate headed downward back into the mantle and presumably to the end of its identity deep in the viscous mantle is made up of jagged edges, various sedimentary deposits, and possibly even seamounts. Here and there along the plate's edge, the process will stall until lateral force becomes sufficiently titanic to overcome the local friction. The results are typically the largest earthquakes. In other words, the ocean bottom disappears in a herky-jerky fashion, in huge chunks up to six or seven hundred miles long and up to three hundred miles deep.

By the mid-1970s, all of these actions were being monitored and analyzed by seismological and other equipment that the great seismotinkerer, John Milne, could not possibly have foreseen. Seismographs, impelled by military needs, were improving to the point where the hypocenters of earthquakes and other information could be pinpointed with almost unerring accuracy. Seismographs were in the process of converting from mechanical devices to electronic ones, and soon computers would be routinely analyzing such information in unprecedented quantities and with previously unimaginable speed. Meanwhile the military provided geophysicists and others with another boon — the Global Positioning System (GPS). Designed to simplify military navigation, it soon became a powerful tool as well for the study of geodesy, the branch of geology that focuses on the size and shape of the planet. Here was a tool that provided details down to a few feet or inches, and could record the slightest slips of faults. The motions of the plates and of parts of the plates could now be monitored and observed continuously and over days, months, and years. Also, since the GPS depends on super-accurate atomic clocks to measure the time of arrival of signals sent from the earth's surface to several satellites, the system came into use timing the precise arrival of seismic waves at seismic observatories. As well, twin lasers could stare at each other across known faults and record (and electronically transmit) the slightest relative motion.

Armed with these new technologies, earth scientists hoped that earthquakes (and volcanic eruptions) would one day be predictable. Much of that hope arose from the new geological model, plate tectonics, that rendered earthquakes now fundamentally understandable.

At least most of them.

Midplate Enigmas

On January 12, 1982, the westernmost station of the Memphis Area Regional Seismic Network detected a Magnitude 1.3 earthquake (otherwise known as a microearthquake) in a seismically inactive plot some 80 miles west of the seismograph and a little north and east of the town of Conway, Arkansas, which itself is a bit north of Little Rock.[3] Within a week, a temporary network of seismic stations had been established in the area just in time to record a Magnitude 4.5 quake on January 21. By October, nine months after the first small quake, a swarm of seventeen thousand other microearthquakes had been recorded, the largest such swarm ever recorded in the central or eastern United States. This oddity was augmented by the fact that all seventeen thousand quakes occurred in an area about three and a half miles square. Nor was there anything especially peculiar about that little area. There were no geological anomalies associated with it, no prominent geological structure. The sources of the quakes were between a half a mile and five miles deep.

The little earthquakes of the Arkansas swarm continued sporadically, quiet periods being interrupted by sudden swarms, bursts of activity well above the overall average. Some bursts appeared to be aftershock sequences of a larger shock, but also foreshock-mainshock-aftershock sequences were seen, as well as bursts that followed no large shock but arose of their own accord. Lots of microseismic information poured in, all baffling. The amount of seismic energy released in the swarm in its first six months was the equivalent of a 4.7 shock — and that exceeded all of the New Madrid seismicity in the previous six years.

The New Madrid area had continued to be the most seismic region of central and eastern America, and it continued to draw the attention of some seismologists — mostly local ones. The main seismological action was, of course, in California, the

Pacific Northwest, and Alaska, where the North American plate was always engaged in its dance of havoc with the Pacific plate. The San Andreas Fault was probably as well known a name as the Los Angeles Dodgers. On the West Coast, a seismologist could expect plenty of action right in the immediate neighborhood. It was there where American seismologists could see at least the far-off glimmering shine from that holy grail — the ability to predict earthquakes.

Midplate earthquakes, not to mention the Arkansas earthquake swarm, were still almost wholly enigmatic, usually explained in rather vague terms as the result of some kind of ancient weaknesses in the crust that are somehow reactivated by stresses generated by the motion of the plates — hardly a precise picture. Virtually every state in central and eastern America experienced earthquakes from time to time, but they tended to be few and far between. The study of midplate quakes was a seismological backwater. Even so, some of the profession remained on the case of the more mysterious and more occasional quakes and swarms east of the Rockies, and particularly New Madrid.

PART THREE

LOOKING BACK,
LOOKING FORWARD

Nature, to be commanded, must be obeyed.

— Francis Bacon

All things have second birth;
The earthquake is not satisfied at once.

— William Wordsworth

8

Rifts, Plumes, and Reservoirs

NOT UNTIL AFTER 1974 was there much likelihood of anyone discovering what sort of forces caused the New Madrid quakes of 1811 and 1812 and kept the region vibrating as one of the most seismic areas on the continent east of the Rocky Mountains. That was the year that St. Louis University, with the support of the United States Geological Survey, installed a network of modern seismographs in the region. Five years later, a second network was installed by Memphis State University (now the University of Memphis) with the help of the Nuclear Regulatory Commission. It had by then dawned on the nuclear industry and its governmental handmaidens that it was probably a poor idea to locate a nuclear power plant in a highly seismic area, for to earthquake-proof such a structure would add millions of dollars to its cost, and there was reason to think that no building could be made that would resist a monster quake — say, a Magnitude 8. (One of the upshots of this collaboration is that there is no nuclear power plant within two hundred miles of New Madrid.) This second network of seismometers was operated by Memphis State's Tennessee Earthquake Information Center —

now called the Center for Earthquake Research and Information (CERI) — under the direction of a young assistant professor named Arch C. Johnston. Johnston had become interested in geology while serving in the Air Force piloting C130E transport planes in long flights over the Andes. Over the next two decades plus, he would emerge as the person who knows more about the New Madrid quakes — past and present — than anyone alive.

By 1980, the New Madrid earthquake zone was being monitored around the clock by instruments sensitive enough to register a Magnitude 1 disturbance. This means that if someone shoved an SUV out of the second story of a parking garage on the outskirts of Memphis, its impact would register on at least a few of the seismometers. In the seven years from 1974 to 1981, more than a thousand quakes greater than Magnitude 1 were recorded in what was called the New Madrid Seismic Zone; of these, residents of the region reported feeling fifty.

Additionally, timing is of crucial importance — if a seismometer makes an error of a tenth of a second in the arrival time of a seismic wave, it creates an error of a mile or two in reckoning a quake's hypocenter. But with the highly sensitive instruments in place, what had seemed to be a random scattering in hypocenters in the region now emerged as fairly distinct linear arrangements. In fact, it appeared that three linked faults were responsible for virtually all the quakes.

One of many questions raised by this map of faults was what caused the faults in the first place. This was, after all, a region located roughly in the middle of the North American plate, a place of supposed stability. And why was this area, with its faults, so much more seismically active than other places in the midcontinent? Was there anything in the still relatively new arena of plate tectonics that could explain such phenomena?[1]

Earlier, in the forties, a member of the Army Corps of Engineers had discovered a crucial fact about the history of the Mississippi River: some eighteen thousand years ago, toward the end of the Pleistocene, the riverbed lay several miles to the west. Over time, in fact, the Mississippi has inhabited so many different riverbeds that, if you could watch it on super-fast-forward from above, it would look like the writhing of a snake. Once the ice retreated back into Canada, the channel eventually moved eastward to its present configuration, in which, just above the place the Ohio joins it, it flows through a narrow pass called Thebes Gap. Could this rearrangement of the landscape have occurred, Johnston wondered, as a result of tectonics — i.e., earthquakes like those of 1811 and 1812?

Plutons, Cratons, and the Reelfoot Rift

When modern seismologists and geophysicists decide to put what might be called a full-court press on a given area, they have many "weapons." Many of these were brought to bear on the Mississippi Embayment, a large U-shaped feature the northern part of which coincides approximately with the New Madrid Seismic Zone, and the southernmost part of which is the bayou country of southern Louisiana. Here, over the eons, river-borne sediments had been deposited above the bedrock, one effect of which was to cover up the three faults that had only now come to light. Thanks in part at least to all of the river's tributaries, the thickness of the sediments increases from north to south, reaching a depth of some fifteen thousand feet.

In the late seventies and early eighties, scientists from the U.S. Geological Survey and from a variety of universities ranging from those nearby (like Memphis State) to as far off as Tel Aviv began the complicated job of "x-raying" the area. Earthquake data from the existing network of instruments were augmented by setting off explosions in carefully chosen sites to get various profiles of the structure of the underlying crust. With the precision timing of the waves' arrivals, showing among other things different speeds of transmission at different depths, the scientists could perceive the layers of different kinds of rock that made up the subsurface structure (and geological history) of the area. They also found the deep-down presence of a graben. In German, the word means a ditch. In geological parlance, a graben is a part of the earth's crust bounded on at least two sides by faults that has been moved downward relative to the adjacent parts of the crust. The graben deep beneath the Mississippi Embayment seemed to be filled with marine sediment. It would appear to have sunk when an underlying rift in the earth's crust gave way. The rift probably began in Precambrian times (earlier than 600 million years ago) but then stopped spreading apart — making it a failed continental rift. About 180 miles long, it is some 50 miles wide and its bottom is at a depth of about 3 miles, compared with its "sides," which lie at less than a mile down. This was quickly named the Reelfoot Rift, underlying as it did much of the New Madrid Seismic Zone and the lake of that name that had come into being through the huge events of 1811 and 1812.

Rifts are where the earth's crust tears apart or at least begins to. For example, a huge rift occurred when what is now South America separated from Africa as the Atlantic Ocean began to form. That rift, still active of course, is now the Mid-Atlantic Ridge. Smaller active rifts are still occurring. The Red Sea lies over such a rift, and the Arabian Peninsula is, as a result, inching apart from Eritrea and Ethiopia on the shore of eastern Africa. What is called the Salton Trough in California is another active one, breaking onto the surface to the south where Baja California has separated from the bulk of Mexico. The Rio Grande, where it flows south out of Colorado and basically splits New Mexico down the middle before wandering southeastward to form the Texas-Mexico border, lies in a rift valley that extends well south into Mexico, making it the second largest rift valley on the planet. There is some disagreement over whether it is still minimally active or truly defunct, a failed rift. In both cases (Salton and Rio Grande) the underlying continental crust has been stretched, thinned, and weakened.

What could be the cause of such rifts? At the Mid-Atlantic Ridge, two plates are being fairly smoothly pushed apart, with the North American plate moving to the west at some infinitely slow pace but with an infinite force. Its leading edge is grinding up against the Pacific plate, which is resistant, holding up the plate's process at least here and there for short periods. Thus, titanic pressures are exerted on the plate and the continent that rides atop it.

The continent of North America consists of two essential parts. First is a large, essentially circular part that is the continent's oldest, strongest, and most rigid crustal rock. This area, called the North American craton, is surrounded on the east, south, and west, as well as by Greenland and a few Arctic islands in the northeast, by what is called coastal plain rock, less rigid and strong, and younger. Except for the seismic activity that results from the North American plate crashing into the Pacific plate, most of the other seismicity occurs around the edges of the craton. It stands to reason that the ancient and superstrong rocks of the craton would be more likely to resist whatever pressure the continent receives as a result of the "logjam" on the West Coast than the weaker newer rocks of the coastal plain. A look at the map shows that New Madrid lies pretty squarely in the middle of a bulge of coastal plain rock that leads from the south into the craton. To hypothesize a rift located in this bulge of less strong coastal plain rock was perfectly reasonable.

In the period when the Reelfoot Rift came to light, other scientists were making magnetic studies of the area as well as measuring variations in gravity. These studies confirmed that there was a rift beneath the upper Mississippi Embayment area, and they also produced evidence that this failed rift had been "re-activated" for some period in the Mesozoic Era, between 100 million and 250 million years ago. The evidence consisted of gravitational and magnetic readings pointing to the existence of intrusions of heavier, metallic rocks — the sort of material that comes from the lower parts of the crust or the mantle below. They lay below the Reelfoot Rift. Such intrusions are called plutons, and the ominous sound of the word matches their typi-cally dark mineral makeup — plenty of iron and magnesium. When they rose up into the crust, they reactivated the Reelfoot Rift for a time and altered the lower crust below the rift.

At the same time, other geologists were finding — through various techniques including the extraction of deep cores from the ground — that much of the deep rock under the New Madrid Seismic Zone was alkalic, or basic, in nature. It was magmatic material from the mantle. At depth it contains a great deal of vol-atile materials like sulfur dioxide, carbon dioxide, and chlorine, and as such volatile-rich magma rises, escaping the pressures of the deep, it cannot hold on to the volatiles, which explode out of the rock.

With all these phenomena now associated with the New Ma-drid Seismic Zone, there was still no ready explanation for the New Madrid quakes of 1811 and 1812. But then the huge Ar-kansas earthquake swarm began in January 1982, with thirty thousand quakes recorded in its first two years. It was, and is, the largest such event ever recorded in the central and eastern United States. Arch Johnston of what was then known as Memphis State University initiated the seismic study of the area in which these swarms were recurring like bursts of machine-gun fire. While the Arkansas swarm was still under way, Johnston suggested what to many geologists seemed a radical explanation, something differ-ent from the prevalent notion of reactivation of a preexisting zone of crustal weakness caused by plate motions. "I suggest," he wrote in the journal *EOS,* "that the swarm accompanied a small intrusion of magmatic material in the shallow Arkansas crust." In other words, an upthrusting jolt of magma under the crust might have relieved the compressional pressure on the crust in that region, allowing a preexisting fault to move in thirty thousand little jerks.

Johnston recalled a much larger swarm that had occurred in the mid-sixties at Matsushiro, Japan, when 700,000 quakes occurred over a two-year period. The source area was only twice the size of that of the Arkansas swarm, and other phenomena — uplift, changes in gravity and magnetic field — had indicated a magmatic intrusion. Elsewhere, intrusive activity was believed to have been responsible for an earthquake at Mammoth Lakes in California in 1980, and it showed some similarities with the Arkansas swarm. It all remained speculative, Johnston said.

Volcanic Encounters of the Third Kind

Most of the tools of seismology have been brought to bear on volcanoes as well as earthquakes; indeed, many of the earliest seismologists were more interested in volcanoes. And of course volcanoes and earthquakes have been seen as related phenomena as far back as Aristotle's time. As with earthquake scientists, most volcanologists find themselves spending most of their time in regions of plate tectonic activity, for here the most common volcanic activity is found. Volcanoes arise from the mid-ocean ridges, and they occur when plates subduct beneath others. Iceland, for example, is a child of the Mid-Atlantic Ridge, as are the Azores and Tristan da Cunha. In June 1783, a volcano known as Laki on Iceland erupted, and its gigantic lava flow continued for eight months. Vast quantities of sulfur dioxide were ejected into the atmosphere, creating what has been estimated to be 100 million tons of sulfuric acid rain — the annual supply of the entire world today. A blue haze spread over Iceland and all Europe, reaching as far as Siberia and North Africa. Temperatures dropped an average of seven degrees. None other than Benjamin Franklin, then ambassador to England, opined that this awful pollution could be the result of "Hecla in Iceland and that other volcano [Laki] that arose out of the sea near that island." This was the first known suggestion of a link between volcanism and widespread atmospheric effects.

On the other end of plates, where they meet, volcanic eruptions like that of Krakatoa are the norm. In the summer of 1883, after a few precursor eruptions, the island of Krakatoa disappeared when its volcano blew in a series of violent eruptions over two days — nothing like the almost low-key, eight-month eruption of Laki. The eruptions that wiped out Krakatoa caused tsunamis that swept away an estimated thirty thousand people on Java and Sumatra. The eruptions were heard as far off as 3,000 miles — one of the loudest sounds ever known in history — and pumice was found floating in the Indian Ocean 2,900 miles away. A green haze settled over the earth, with sunsets red enough to cause fire departments to be called out. Almost seventy years earlier, in 1815, the far greater eruption of Tambora, a volcano on the island of Sumbawa also in Indonesia, filled the global atmosphere with pollutants. World temperatures dropped precipitously, bringing about what New Englanders called The Year Without a Summer. It snowed in July in New England, and across the Atlantic calamitous crop failures and starvation in Europe and elsewhere precipitated political revolutions.

Such are the gifts of the Ring of Fire.

A third type of volcano exists — those that erupt in the middle of a plate. Yellowstone National Park is the creation of such volcanism. After a long series of relative modest eruptions, some two million years ago three cycles of huge eruptions occurred.

The first produced about 2,500 cubic kilometers of ejecta and a caldera of some thousand square miles. Only the "fossil" remnants of this crater now exist in the park. The second cycle of eruptions was far smaller, producing the Island Park caldera some seventeen miles across. The third cycle created the Yellowstone caldera, which measures about thirty by fifty miles. The subterranean seething went relatively quiescent, only to awake again six hundred thousand years ago in a massive eruption that spread volcanic ash over every state west of the Mississippi except Washington and Oregon (which were upwind). Compared to this mighty reconstruction of the landscape, the eruption of Mount Saint Helens in 1980 was about as exciting as a good show of fireworks on the Fourth of July.

Another example of a continental midplate volcano is Valle Grande in northern New Mexico. Here, 1.12 million years ago, a volcano that would become the Jemez Mountains blew its top and created a caldera of 180 square miles, one of the largest remaining calderas in the world. In the eruption, layers of volcanic tuff blanketed the landscape, creating the white cliffs in today's Bandelier National Monument where the ancestors of several of today's Pueblo Indian tribes created a honeycomb of dwelling places.

One of the best examples of midplate volcanism lies in the Pacific Ocean on one of the great tourist destinations of that great body of water. Hawaii and other similar archipelagos owe their existence to what Tuzo Wilson suggested early on (and most geologists now agree) were gigantic hot spots — sites where for some reason the earth's crust periodically opens up to permit magma to explode through the crust and create an island. While the plate moves overhead, the hot spot remains in place. One hot spot currently underlies Hawaii, the Big Island. There, four volcanic structures rise into the clouds, the most famous of which is Mauna Loa, the most active one, which erupts periodically, predictably, and nonexplosively, making it one of the safest active volcanoes to approach in the world. Indeed, if you look at a map of the Hawaiian Islands and the Leeward Islands that lie beyond to the west-northwest, you can visualize the motion of the Pacific plate over the eons.

Proceeding west-northwestward from the Big Island are Maui, Molokai, Oahu, and Kauai. Each is older than the one to its southeast. Beyond Kauai, the Leeward Islands are worn down to a string of nubbins, home to seabirds now and not much else. Beyond them, in the same northwesterly direction, are a few tiny island pinnacles and underwater seamounts, known as the Emperor Seamounts, which abruptly turn toward Russia. At that point, the seamounts are some seventy-eight million years old. Hawaii itself is only six million. So over a period of time beginning a bit before the end of the Cretaceous about sixty-five million years ago, when dinosaurs still roamed parts of the world, magma from the hot spot now beneath Hawaii periodically punched through the crust of the earth, creating island after island as the Pacific plate made its stately way north-northwest. Then forty-six million years ago something unknown caused the plate to change direction and head west-northwest — through all of this long history moving at a rate of about half a foot a year.

A question arose: Why would such a hot spot remain in the same place over the eons? In the 1970s a physicist at Princeton University, Jason Morgan, suggested the existence of magma fountains originating deep below the surface, great plumes of magma that rise vertically. Such plumes, Morgan suggested, occur as singletons or small clusters here and there beneath the plates themselves. He proposed that there were some twenty such plumes, including those below Hawaii and Yellowstone. In addition, mantle plumes appeared to occur in long linear regimentation at the mid-ocean ridges, most notably near Iceland. They rose, presumably, from near the boundary between the lower mantle and the liquid core of the earth, about halfway to the center of the earth. The rising molten material in these plumes is basic in composition (as opposed to acidic), dominated by heavy metals and the volatiles.

Until recently this has been widely accepted by the geological community as *the* explanation of the hot spots that definitely do reside at so many points around the globe. But in the 1990s, a handful of geologists began to doubt that such plumes existed. In 1996 a British seismologist, Gillian Foulger, laid a network of seismographs across Iceland that acted in the manner of ultrasound, the technique undergone by many a pregnant woman that permits doctors and her to observe the fetus — but in this case of Mother Earth. She began to get puzzling data. To begin with, it should be recalled that molten magma is less dense than solid rock and thus slows seismic waves down. Instead of the long thin mantle plume she and her team expected to see originating from the core-mantle boundary some 1,700 miles down, there appeared to be only a big blob of magma some 250 miles down, a big reservoir.

Foulger found support in the person of geophysicist Don Anderson of the California Institute of Technology, who was saying that mantle plumes do not exist. Why, he asked, would anyone think that a tectonic plate some seven thousand miles across would not be internally stretched and, instead, remain unbroken? And if there was a break in the crust, why wouldn't molten rock flow upward and break through the surface?

In the first place, Anderson pointed out, when Morgan developed his notion of mantle plumes, it was generally held that the mantle was of uniform composition and not all that hot. By the mid-nineties, it was known that the mantle varies quite widely in both composition and temperature, that some island chains thought to have been created by plumes actually vary widely and randomly in age, and that some of the hot spots themselves actually move relative to each other at about the same rate that the plates move — a few inches a year.

One early clue in the original mantle plume theory was that magma from the mantle was rich in a form of helium that in Jason Morgan's time was believed to have been left over from the big bang and could have survived on earth only in a deep near-core reservoir. But that form of helium had since been found to exist in a large number of rock crystals with high melting points that are common in the upper mantle, making mantle plumes unnecessary. And the renegade Anderson also pointed out that physics suggests that long narrow mantle plumes simply could not exist — the pressure at the earth's core is simply too high to permit narrow plumes to form and rise. Instead, to attain the mass needed to rise, a blob of magma would have to be some three thousand miles across, and if such a blob did rise up and break through the crust, the lava floodplain that resulted would be larger than anything ever seen on the planet.

Suppose, the renegades say, using the case of Hawaii as an example, that a crack in the Pacific plate began to open up off the Russian coast and began to propagate to the south and east across the Pacific. As the crack opened, magma in a shallow reservoir in the mantle would rise up through the crack and form islands that would then proceed to erode. The oldest ones — near Russia — would be mere seamounts, while the actively volcanic island of Hawaii would lie at the most recent site of the cracking.

The renegades are far from carrying the day, however. Some seismologists, "refocusing" seismic data by means of computer models, have found that mantle plumes do exist in at least seven places, and the phenomenon of crustal faults or cracks opening up over time simply does not explain enough of the earth's geopolitical metabolism to throw mantle plumes out the window. The debate will continue and ultimately be resolved, but in the meantime no one is in much doubt that, whatever the cause and the mechanisms involved, hot spots exist around the globe and, whether they are utterly stationary or minimally mobile, they can occasionally give rise to midplate volcanoes. And this brings us back to Missouri.

A Magnificent Failure

With all this in mind, we can look anew at New Madrid and the Reelfoot Rift, which, like many if not all such failed rifts, appeared from geophysical measurements to have its lower crust altered by the intrusion of material rising up from the mantle. This feature suggests to the authors that at some time a mantle plume at great depth or alternatively a magma reservoir at less depth caused a partial melting, with the subsequent intrusion of basaltic magma into the lower continental crust.

The notion thus arises of what might be thought of as a failed volcano, certainly a magnificent failure — the action of a mantle plume or other source that did not quite manage to punch all the way through the crust and become a volcano. Could this explain the New Madrid quakes of 1811 and 1812? For those quakes exhibited a number of unusual features that can be explained most efficiently by the existence of a mantle plume or reservoir beneath the New Madrid Seismic Zone. First of all, the sounds of the New Madrid quakes — the long roaring sounds — were heard as far away as the East Coast. Earthquakes do not typically roar. Many people smelled sulfur, and sulfur dioxide fumes permeated and darkened the atmosphere in the neighborhood of New Madrid. These fumes are not a typical earthquake accompaniment. There were three major quakes, all of comparable magnitude. Aftershocks went on for several years. None of these features are associated with the strain release from interplate earthquakes. All of them, however, are associated with

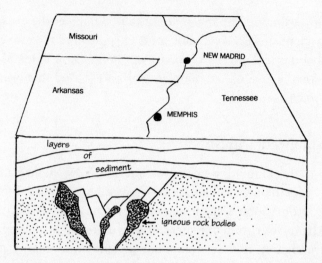

Figure 12. A cutaway shows the Reelfoot Rift, where the ground slumped downward and later filled with marine sediment. Over the eons, sediment continued to be deposited in layers. The rift was then intruded by igneous rock bodies, indicated by dark-spotted areas. Millennia of river sediments have buried the rift. Jake Page, derived from Arch C. Johnston, "The Enigma of the New Madrid Quakes of 1811–1812," *Annual Review of Earth and Planetary Science*, 24, pp. 339–384, 1996.

large volcanic eruptions: sounds heard thousands of miles away; gas emissions that pollute the atmosphere; and more than one major eruption with an eruption sequence that lasts over several years.

Whether this modest proposal actually explains those quakes remains to be seen, and it leaves several big questions unanswered. For example, did the existence of the Reelfoot Rift provide a weakened area into which some rising magma opportunistically diverted? Or did the rising magma cause the rift? Arch Johnston of the Center for Earthquake Research and Information at the University of Memphis does not think much of this proposal, and for several reasons, both of which are critical in his view. He says that if there were a hot spot below the Reelfoot Rift and the network of faults above it, there would be a more severe temperature change between the lower and upper rocks down there. Such is the case, for example, at Yellowstone, where Old Faithful faithfully expels water and heat with the regularity of a metronome. Furthermore, if there were a great blob of magma down there, the crust would be thinner — perhaps about ten miles thick, Johnston says, rather than twenty-five, which is its local thickness. Perhaps most telling is that if a hot spot were lurking below, the entire Mississippi Embayment would be rising, however slightly. In fact, it is falling. For Johnston, the New Madrid quakes of 1811 and 1812 still call up the words of Winston Churchill, who, when describing Russia long ago, said, "It is a riddle wrapped in a mystery inside an enigma."

Johnston's concerns in the past decade or so have been less theoretical, less about the ultimate cause of the New Madrid quakes, and of more immediate, practical consequence: Will the New Madrid quakes happen again? And if so, when? Can such earthquakes, these strange seizures that take place around the world where the crust is essentially stable, be forecast?

9

The Art of Prediction

IN EARLY MARCH OF 2003, an earthquake began in the Pacific Northwest unbeknownst to everyone in that region except for a handful of seismologists. (Yes, the word is "began." It was still under way in late April.) It was what is called a slow quake. Slow quakes appear to occur in this region every fourteen and a half months or so. They have only recently come to the attention of scientists because they take place below the sensitivity threshold of most seismographs in use. They were first detected about a decade ago by global positioning satellite technology, which can now track even the tiniest movements of the earth's crust.

The slow quakes in the Northwest are thought to work like this: offshore in the Pacific Northwest, the Juan de Fuca oceanic plate bumps into and dives under the continental plate upon which sits the state of Washington, among other pieces of real estate. GPS monitors have found that, as expected, the land upon which they rest is inching fairly steadily to the northwest. But every fourteen and a half months (plus or minus a few days or even a month) the monitors are found to be moving to the southeast. Soon, after a few weeks, the northwestward motion resumes.

What may be going on is that deep down where the plates are hot and plastic, they slide relatively smoothly past each other. Even so, they can occasionally become hung up, it seems, and stress begins to build up. Eventually it is released, causing the silent, almost unnoticeable slow quake. A slow quake apparently can release as much energy as a major quake closer to the surface releases in seconds, but since a slow quake takes place over weeks if not months, it is of itself harmless. It would, in fact, be altogether harmless except for the fact that higher up, nearer the surface of the earth where the plates are less plastic, they also get hung up and are building up a tremendous amount of stress that will one day be released in what many seismologists expect to be a calamitous megathrust earthquake — the Northwest's Big One. Researchers now wonder if the recently discovered, regularly occurring slow quakes could be precursors. Could one of these subtle, deep, silent quakes be the straw that breaks the camel's back — that sets off a major quake when the upper crust is nearing the breaking point? Scientists have found that over the past ten years, eight slow quakes have taken place in the region. They have also found that similar slow quakes that registered in both Chile and Japan have preceded major quakes. The question arises: Should officials in the Pacific Northwest start preparing the citizenry for a major quake to strike sometime in the early summer of 2004, and failing that, the late summer of 2005? And every fourteen and half months thereafter?

Probably not. To begin with, slow quakes have only been known for the past ten years.

And, while everyone loves to hear about regular cycles associated with anything from sunspots to stock market behavior, there is no real proof that slow quakes trigger big bad quakes. Indeed, earthquakes would appear for the most part to be far more complicated than even the climate. And climatologists at least have the benefit of being able to see what it is that they are try- ing to predict. In fact, some people in the seismological world will tell you that there simply are too many variables, too many unknowns, too many physical complexities to ever be able to pinpoint an earthquake's occurrence closely enough to make such measures as evacuation practical. No one, after all, knows which slow quake (if any) might presage a major quake. And it would not take too many unneeded evacuations from Seattle, say — every fourteen and a half months the citizenry packs up and heads for the hills — before the citizens, their insurance compa- nies, and their elected representatives would see to it that seis- mologists were confined to some kind of detention center.

The above exaggeration is designed to point up the difficulties — scientific, yes, but also societal and ethical — that face those who would predict earthquakes, especially those who would pin- point them so finely as to forecast the date and place of occur- rence and the expected magnitude. The quest to predict quakes extends, of course, to New Madrid and other midplate quakes, though most predictive efforts have been made on quakes that occur on the edges of plates, as on the West Coast of America.

Other kinds of earthquake predictions (in fact, they are not technically predictions but forecasts) are easier to make, and less given to creating unnecessary alarms and excursions. For exam- ple, where the Pacific plate grinds past the Alaskan plate, large sections have been sprung while others remain fixed. It is not all that difficult to forecast that those places that have long re- mained fixed are almost surely going to suffer a huge slip — and almost certainly before those sections that have already slipped. That is a forecast. A prediction (as seismologists use the word) would call for more specifics, such as when, and how big a quake. Also, given the vast amount of information that has been amassed — in this country particularly in California — seismologists began to provide forecasts of yet a different kind.

Beginning in the 1960s, attempts to predict earthquakes grew rapidly in such seismically active places as Japan, China, Russia, and the United States. It was a decent, even a highly worthwhile goal then and still is. Earthquakes in the United States rarely cause the enormous loss of life seen in some other highly seismic countries. It has been estimated that heat prostration in Chicago in one particularly bad summer around the turn of this century killed more people than all the U.S. earthquakes since 1906. Parts of the United States and particularly California are well prepared for such events, while many parts of the world are not. Turkey, for example, is struck by large quakes with a regularity that sometimes seems like the work of a malevolent cosmic drummer. Over the years Turkey has instituted building codes and other preventive measures, but much of the country that lies along the North Anatolian Fault is characterized by old towns with old buildings that are distinctly not earthquake-proof. When one strikes, entire villages can be flattened with enormous loss of life.

A quake elsewhere suggests the political turmoil that can result. In May 2003, a 7.1 quake struck a small Algerian city on the Mediterranean Sea and buildings collapsed like so many card houses. More than two thousand people perished, with another ten thousand injured. More thousands were homeless, and the city was without basic services. Riots against the government broke out, a reminder that in countries where the traditions and structures of civic society are fragile, an earthquake can bring on seismic problems for governments.

In all, it is estimated that hundreds of thousands of people worldwide die in earthquakes every year. This number could rise significantly in the future, but not because large earthquakes are any more likely to occur. The problem is that cities, particularly third-world cities, are growing in population at a tremendous rate, and many of the world's megacities — such as Istanbul, Mexico City, and Jakarta — lie on or within a hundred or so miles of a plate boundary or major quake zone. Characterized by vast shantytowns and buildings not designed with earthquakes in mind, and reaching the tens of millions in population, such cities could see millions die in a single major quake.

One early form of predicting quakes focused on forecasting the seismic intensity of an earthquake that would probably happen in a given general locality. Such a prediction, though not telling when it might happen, provides cautionary advice as to the placement of important long-term facilities like dams, hospitals, schools, power plants, and so forth. And such predictions are given in the form of probabilities — for example, a quake of between Magnitudes 5 and 6 may — probably — occur in the neighborhood of XYZ City between the years 2005 and 2015.

The problem with such a forecast is that it would for obvious reasons have been made about a zone that already was known to be seismic, probably highly seismic. Thus the odds are not zero that an earthquake will occur in the given period in the given neighborhood. And if such a quake does occur, it does not decisively prove that the method by which the forecast was made was correct. One needs a large number of quakes that fall within the probabilistic parameters before the method can be taken as useful. Of course, if the forecast quake does not occur within the ranges specified, the method can be taken as flawed in at least some important particular.

To have a useful prediction then, scientists have agreed on the need for four criteria. These are the criteria used, for example, by a California panel set up to evaluate earthquake predictions for the State Office of Emergency Services in 1975 and by a similar national panel — the U.S. National Earthquake Prediction Evaluation Council — set up later to advise the U.S. Geological Survey. A prediction must include (1) the period in which the quake will occur, (2) the area of its occurrence, (3) the range of magnitude expected, and (4) the odds that an earthquake like that predicted could occur without any special evidence but by chance alone.

Some Costs of Prediction

Evaluation panels are of great importance since the reliability of predictions — and vanishingly few have proven reliable yet — is paramount. Predictions can unleash a great deal of activity and expense. Try, for example, to sell your house if you live in an area where an earthquake has been predicted to occur in the next few months. Property values in such situations are certain to drop and, with them, property tax revenues. Insurance either becomes harder to find or more expensive or both. Mortgages are hard to obtain. People avoid high-risk parts of nearby cities, lowering commercial activity. Employment opportunities drop, and except for expenditures on preparedness, public services may be reduced. This short list of socioeconomic impacts does not, of course, include anxiety and other personal emotional costs to both adults and children, adding potentially to medical and pharmaceutical costs in the area. On the other hand, to be able to predict earthquakes with some degree of accuracy would be a great boon to the world — thousands of people die in earthquakes every year, so even a probabilistic prediction that could be counted on would be a blessing.

Indeed, probabilistic predictions remain for many professionals the holy grail of seismology, and some few pursue even more pinpointed predictions — all this with ever increasingly sensitive technologies, eternally springing hope, and a fervor that has burned since the time of John Milne.

Biting Pigs and Crying Sea Lions

A few pinpoint predictions have come true. In early February 1975, in the Liaoning Province of China, just north of the Korean peninsula, pigs began biting one another and trying to run up the walls. Deer fled, snakes came out of hibernation, and cows got in fights. Not only that, but groundwater levels rose and fell, radon gas levels changed drastically, and the ground tilted. A month before the animals went nutty, the provincial government knew something was up and its Earthquake Research Branch predicted a major quake the following month (February) or the next. Only six years earlier another place in the province had been struck by a major quake, so the province was now laced with seismographs and other monitoring devices. By late January, in some places, water stopped flowing altogether, the electrical conductivity of the ground itself changed, and by February 1 a swarm of tremors or foreshocks began and increased in frequency and magnitude to a peak on February 4. On the morning of the fourth, the provincial government issued a warning and began an emergency evacuation of the city of Haicheng. That night, at 7:36 P.M., a major quake hit the city, Magnitude 7.3, and an incalculable amount of loss of life was averted. It was the first time ever that a major quake was forecast so precisely . . . and it was the last.

Animals doing crazy things, radon gas levels changing, wells losing water, severe weather conditions, volcanic activity, the gravitational pull of the moon, the sun, and the planets, clusters of foreshocks, slow quakes, even winter snow and the melting of glaciers — all have been put forth as possible triggers for the failure of faults. None has proved at all conclusive. There remains no viable theory of short-term events that lead up to earthquakes. In July 1976, not quite a year and a half after the triumph of earthquake prediction at Haicheng, another one occurred at Tangshan, also in northeastern China. At 3:42 in the morning it reduced the city of a million to rubble and abruptly ended the lives of nearly a quarter of the city's population. There had been changes in groundwater levels, electromagnetic oddities, and strange behavior by domestic animals, but there had been no foreshocks. And there was no prediction.

The question often arises, particularly among laymen, about strange animal behavior that in so many accounts precedes a quake. When it occurs, it usually starts a few hours or a few days before the quake. It has been documented in many instances in China and elsewhere around the world in such places as Italy and Japan. After the San Francisco quake of 1906, the commission founded to explore the quake reported that several farmers had noticed their horses snorting and whinnying before the quake was felt, and in one case, cows stampeded. Most scientists consider such stories to be at best pleasant distractions and at worst sheer nonsense. Most scientists are at pains to eschew anecdotal events. In science, an anecdote is an unsubstantiated event witnessed or described after the fact usually by an untrained observer.

It is possible to quibble about this fastidious aversion on the part of scientists, asking for example how many similar anecdotes are required before they qualify as a pattern. (A pattern in events is a very good thing in science.) Furthermore, earthquake scientists have long been in the business of asking untrained observers about the extent of a quake's damage in order to judge its intensity on the Modified Mercalli Intensity Scale and others. Not only that, it could be argued, but most people who keep livestock such as horses and pigs are extremely alert to abrupt changes in their behavior, making them qualified observers of livestock behavior if not earthquakes. Finally, modern seismologists have made very good use of the observations citizens made at the time of the New Madrid earthquakes of 1811–1812.

Still, reports of oddball animal behavior before an earthquake are taken to be incidental at best. In the first place, they don't clearly point to any knowledge about what exactly goes on at the hypocenter that triggers the quake. Nor could they be used as a standard predictor of earthquakes since not all earthquakes are preceded by them. More to the point, however, is the likelihood that when such behavior occurs it is because the animals sense a big quake's foreshocks. And not all quakes preceded by foreshocks are "predicted" by cows doing silly things in response. So such accounts are given little significance by the profession, which would rather look for such precursors as foreshock patterns with increasingly sophisticated measuring devices rather than rest their case on the sensorium of a cow.

Even so, some seismologists, and especially the Japanese, still keep an eye out for such behavior. In 1995, in the port city of Kobe, a particularly damaging earthquake occurred, and there were reports of fish swimming in uncharacteristic ways in the bay, and of sea lions crying and not eating. In the local zoo, hippopotami sank into their pools and refused to resurface, while crocodiles became extremely violent. Investigators at nearby Osaka University suggested that the strange animal behavior was a result of a different precursor than foreshocks: changes in the electric field. They have coined a scientific-sounding phrase for the phenomenon with its own acronym, SAAB, which stands for Seismically induced Anomalous Animal Behavior.

Precursors and Prediction Programs

When the United States undertook an official earthquake prediction program in the 1970s, under the direction of the U.S. Geological Survey but including a host of participating university scientists, one often heard that accurate and useful earthquake predictions were "imminent." Hopes were high, what with the leap in sensitivity of modern seismographic equipment and the rapid accumulation of data. Since then, more modest projections have been the norm. Like other such studies throughout the world, the U.S. program sought identifiable precursors to major quakes, chiefly in measurements of seismicity and crustal movement and strain. Today, along the San Andreas Fault system — the main focus of seismic research still — hundreds of highly sensitive instruments feed elaborate computing facilities with micro- and macroseismic information, detecting, counting, and determining the location of every tremor, however slight, and looking for the swarms of small earthquakes, or foreshocks, that sometimes precede a big one. Lasers on both sides of the faults stare at each other, attuned to detect any relative movement of the Pacific and North American plates. Tilt meters measure local changes in ground surface, strain meters in drill holes measure vertical strain, and other instruments track water levels in wells, radon gas emissions, and changes in such things as geomagnetism and geoelectricity. Yet other instruments measure large-scale subsidence and uplift of the earth's surface.

Serendipity sometimes plays a role in the advancement of science. For some time before the Loma Prieta quake struck northern California in 1989, among other things collapsing the elevated highway in Oakland and a section of the San Francisco–Oakland Bay Bridge, an independent group of scientists from Stanford was studying low-frequency electromagnetic radiation only five miles from the epicenter of the quake. About a month before the quake, they recorded an increase in background noise for the lowest frequencies being monitored (0.05 to 0.2 cycles per second). Twelve days before the quake, the background noise rose for the entire ultra-low-frequency (ULF) range — those sounds that lie well below the range of human hearing. Then, spectacularly, three hours before the event, the background noise rose by a factor of thirty — a level the scientists had not ever seen in two years of monitoring.

Was this huge spike in ULF radiation a happenstance or a real precursor? No one at the time had any idea what low-frequency electromagnetic radiation might have to do with earthquakes, and the investigators suggested perhaps there was a piezoelectric effect (electricity emitted from a mineral because of intense pressure on it) from the vast energy release of the earthquake. Russian scientists went back over their files and found that in the Republic of Georgia the Magnitude 6.9 Spitak earthquake on December 8, 1988, had been accompanied by the same sort of ULF signals. On the other hand, the 6.7 Northridge quake in the Los Angeles area on January 17, 1994, showed no ULF precursors. An array of ULF receivers was quickly distributed along the San Andreas Fault system and seismologists await another quake. Meanwhile, most seismologists hold out little hope for a meaningful precursor to arise from ULF electromagnetic radiation.

Another potential precursor was suggested by Russian seismologists who had noted that seismic velocities sometimes changed in anomalous ways before earthquakes. Columbia University scientists recalled this when they began measuring seismic velocities in New York's Adirondack Mountains in 1973. They were setting off small explosives to see how long it took for the seismic waves to reach a nearby receiver, and on August 1 saw anomalous changes similar to those reported by the Russians. They predicted a quake of Magnitude 2.5 to 3.0 in the next three days and — voilà! — it occurred on August 3, Magnitude 2.6.

A fairly accurate precursor is the construction of a big new reservoir. In 1975 in California, a seismograph network was installed to monitor the effects of the additional weight on the earth's crust resulting from construction of a large dam and the filling up of the reservoir behind it. As the reservoir fills, it was assumed, the extra weight of the water can put sufficient strain on the earth's crust to make the land subside either slowly or, in some cases, abruptly via earthquakes. In June, microseismic activity increased significantly and the scientists warned the Department of Water Resources to expect an earthquake. On August 1, a Magnitude 5.9 quake shook the area. (The Chinese have consulted earthquake specialists here and abroad, expecting that earthquakes will occur during the few months it would take for the waters of the Yangtze River to fill the hundred-odd-mile-long reservoir behind the massive dam they have built.)

The following year, in June 1976, the Geological Survey noted an increase in creep (small movements) along the Calaveras Fault that lies east of San Jose, California. They predicted a quake would occur there within a three-month span centered on January 1, 1977. The quake, Magnitude 3.2, occurred on December 6.

Other than these few almost serendipitous examples, predictions — especially of larger quakes — have been disappointing, and none more disappointing than the great scheme for a tiny town in Southern California called Parkfield upon which descended in the 1980s what might be thought of as the full panoply of earthquake prediction technology and dreams. About halfway between Los Angeles and San Francisco, Parkfield is in a region of rolling open land and home to some twenty souls — chiefly ranchers. It is remote and lies upon a visible and well-understood fifteen-mile section of the San Andreas Fault. Beginning in 1887, the University of California had monitored it for earthquakes, noting that moderately large (5.5 to 6.0) quakes occurred nearby in 1901, 1922, 1934, and 1966. Furthermore, similar quakes evidently occurred there in 1857 and 1881. Simple arithmetic shows a fairly regularly occurring incidence of these moderate quakes — about every twenty-two years. The exception was the 1934 quake, which came too soon, after a mere twelve years, with the next one taking place after a thirty-two-year interval. On the other hand, some pointed out an over-all consistency: there were exactly forty-four years between the 1922 and 1966 quakes. The 1934 quake was thus the aberration in a wondrous sequence of nearly perfect regularity. It also appeared, from records of some of the quakes, that they were preceded by clusters of small foreshocks.

In any event, in 1984, two seismologists predicted that a Magnitude 5.5 to 6.0 earthquake would occur at Parkfield in 1988 (plus or minus four years), and this "experiment" earned the official seal of approval from the National Earthquake Prediction Evaluation Council. Parkfield was soon bristling with all manner of seismometers, strain meters, and other instruments. Ten-foot-long devices were laid across the San Andreas Fault to measure creep; lasers measured the distance across the fault. Other devices were sunk into boreholes to measure changes in the volume of rock, the result of building pressure. Such measurements were deemed necessary since, in the 1966 quake, some precursor or another had broken an irrigation pipeline that lay across the fault some nine hours before the actual quake. Over the next few years researchers perceived small signs — small rumbles and other indicators — that suggested that the Parkfield leg of the San Andreas Fault was getting ready to slip. In 1986, the magazine *Science* published a note titled "Parkfield Earthquake Looks to Be on Schedule." But these were all false alarms.

By 1992, no moderately sized quake had happened in Parkfield and the experiment was deemed a failure. This was, of course, a terrible disappointment, for if there had ever been a likely place (with its associated record of quakes) for a prediction, it was Parkfield. A reprieve of sorts came when a computer model of the San Andreas Fault in the Parkfield vicinity suggested that the Parkfield quakes had — all along — been the prolonged result of the huge Fort Tejon quake of 1857 that had rearranged some two hundred miles of the fault from Parkfield to the south. The model suggested that the interval between quakes should be getting longer with the passage of time. The thirty-two-year interval from the 1934 quake to the next one looked, in this light, more like a proper pattern than a fluke. But as the years went by, Parkfield remained largely inert. The ranchers went about their affairs uninterrupted. Questions were raised if some of the Parkfield events in history had actually taken place near Parkfield or if they had in fact been the result of slippage of other fault systems that are adjacent to the Parkfield fault. By the year 2003, seismologists still watched their instruments. Ever hopeful about so promising a site, however, Stanford University and the U.S. Geological Survey hatched a new scheme. By mid-2004, they expected to lower seismic instruments down a two-and-a-half-mile-deep hole drilled smack-dab into the region of the Parkfield fault most given to microearthquakes. Even so, few seismologists regard precision earthquake prediction with an exuberant flowering of optimism. Humility is in.

It is generally conceded now that fault zones do not operate as if timed by helpful metronomes, but most seismologists also believe that the average rate of earthquake occurrence on a given fault corresponds pretty well with the long-term rate of stress, or slip, applied to the fault. The Seattle Fault, for example, which slips at maybe a thirty-second of an inch each year, is likely to be the scene of a large quake every five millennia, while the San Andreas Fault, which slips at half an inch annually, produces large quakes somewhere along its length every few hundred years. On the other hand, many seismologists believe that earthquakes strike essentially at random, and have no notable influence on where or when the next one will occur, except in the broadest sense, and this, if true, makes any kind of precision prediction all the more unlikely. Even so, there are those who still see possibilities.

One of these is Ross S. Stein of the U.S. Geological Survey's Earthquake Hazards Team in Menlo Park, California, who has suggested that earthquakes may, in a sense, hold conversations with one another — conversations that gainsay the notion that earthquakes occur randomly. In fact, one third of all earthquake tremors occur not at random but in clusters in both time and space. They are called aftershocks, and as long ago as 1894 John Milne's close associate, Fusakichi Omori, made observations of them that were then quantified into Omori's Law, which holds that aftershocks occur in great abundance right after a mainshock. After ten days, the rate of aftershocks has dropped to 10 percent of their initial rate. After a hundred days, the rate drops to 1 percent. And so on. Stein and colleagues began looking for similar patterns in the history of earthquakes along the San Andreas Fault and developed a hypothesis called stress triggering.

Its basis is the realization that when an earthquake relieves stress on a fault, the stress doesn't just go away, evaporate, disappear. It moves elsewhere, down the fault, into nearby sites where it can cause tremors. Indeed, the stress appears able to jump from the original fault to others. Stein and his colleagues found that on the day after a Magnitude 7.3 event, the chance of another big shock occurring within some sixty miles is nearly 67 percent, and this is about twenty thousand times the likelihood on any other day.

Taking their hypothesis to Turkey, they calculated where the stress from previous earthquakes had risen, and suggested that a 12 percent chance existed that a Magnitude 7 (or larger) quake would strike the city of Izmit between 1997 and 2027. The quake hit in August of 1999, a Magnitude 7.4 shock that devastated the city. Some twenty-five thousand people died and property damage was in the billions. Five months later in November, a Magnitude 7.1 quake struck some sixty miles to the east of Izmit, near the town of Duzce. One of Stein's collaborators, Aykut Barka of Istanbul Technical University, had calculated the increase in stress from the Izmit quake and published his findings in September. This led officials to close down the school buildings in the town of Duzce, and many of them were flattened by the November quake.

These nearby "conversations" could conceivably turn out to hold a key that will help unlock the secret of earthquake timing and place. But evidence also exists for long-distance conversations. Three major quakes in the western United States (including Alaska) in the past decade are known to have caused swarms of tremors thousands of miles away. Two in Southern California (the Landers earthquake of 1992 and the Hector Mine quake of 1999) and the huge Magnitude 7.9 shock on Alaska's Denali Fault in 2002 caused the magma deep below the geothermal fields in Wyoming's Yellowstone National Park to rumble louder than usual. One of several thoughts is that the seismic waves from big quakes disturb gas bubbles within the magma. The answer to this and many seismological puzzles remains "more tremors, more data."

Yet another potential precursor came (literally) to light in the early years of this century. Stories abounded of strange light effects preceding earthquakes. The philosopher Immanuel Kant noted that the Lisbon quake of 1755 was preceded by violent lightning. The young writer Pierce noted two huge columns of light a few days ahead of the December 16 quake at New Madrid. Floating balls of light were filmed and shown on Turkish television the night before the devastating earthquake that hit the city of Izmit in Turkey. While light effects are generally considered real phenomena associated with at least some quakes, no one had much of an idea what could cause them. But in 2002

Friedemann Freund, a physicist at San Jose State University in California, published a theory that he hopes explains not just the light effects but also a NASA finding from satellite surveillance that found infrared effects in the region near the time of earthquakes. Freund suggests that because of immense pressure before a quake, rock will deform, breaking chemical bonds and electric charges when protons are emitted. These collect, Freund believes, on the surface and create a sufficiently large electric field to ionize the air, causing light resembling a miniature aurora borealis, as well as heat.

Other scientists have technical troubles believing Freund's theory, even though he has produced these effects by crushing rock in his laboratory, and even despite NASA's findings of infrared effects before some quakes. One major problem is that the earth's crust contains a lot of water, an excellent conductor of electricity, and it should essentially short-circuit any underground development of electric charge.

Were Freund's theories to work in the real world of earthquakes, we might well see earthquake predictions being helped along by satellite imagery, but to be truly useful, the theory will have to be like all other precursors that earthquake scientists seek — a phenomenon that is common to most if not all earthquakes.

Whether these relatively new hypotheses of stress triggering and light effects, and others that will no doubt arise, will one day provide a consistent means for predicting the likely time and place of large quakes like the one at Izmit and its follow-up quake near Duzce remains to be seen. And once again, these and other faults at the edges of continental plates are far better understood than whatever underlies midplate earthquake zones, like the New Madrid Seismic Zone. Are midplate earthquakes one-shot affairs, or are they likely to recur? Is there any way of estimating when they are coming?

On April 29, 2003, at about four in the morning local time, a quake deemed to be Magnitude 4.9 struck in Alabama ten miles from Fort Payne in the northeast corner of the state near Tennessee. The quake hit in the southern part of the eastern Tennessee earthquake zone, one of the most active in the entire Southeast but not known for large quakes. The Fort Payne quake, with a hypocenter some nine miles deep, cracked foundations, shook pictures from walls, and roused a large number of people from their sleep in a region stretching as far east as Atlanta. Otherwise, there was little damage to property or people. Here and there, trailers slipped off their foundations, and people reported hearing a low sound like rolling thunder — a tornado some thought at first till they felt the quake as a shuddering of the earth. Some others thought it was an airplane or a bomb.

The quake was felt throughout the South, in Alabama, Georgia, Tennessee, Kentucky, the Carolinas, and Mississippi. The region has experienced a number of smaller quakes over the past century and one other recent 4.9 quake in 1997, in southern Alabama (not in the eastern Tennessee seismic zone). While not a big deal, the Fort Payne quake reminded the world that such midplate quakes and the seismic zones they occur in are among the least understood. No one predicted the Alabama quake: What of the Big One in New Madrid? Will that recur? If so, when?

Some of the world's biggest earthquakes, the New Madrid quakes included, have taken place where the crust of the earth is at least relatively stable, but such quakes are infrequent and account for less than 5 percent of all the seismic energy released by the earth's seizures. Until recently no one knew if these midplate quakes were random one-shot events or the sort of thing that could recur. In the 1980s, a great deal of scientific detective work needed to be done on the world's midplate quakes before anything like an answer about possible recurrence at New Madrid would emerge.

In the mid-1980s, Arch Johnston attended a workshop that included some of the country's most eminent seismologists, and none of them, when asked, could identify the largest ones ever recorded in stable continental crust. This triggered a long-term study by Johnston and his colleagues that was the first attempt to understand all such quakes known to have occurred throughout the world.[1] Plenty of good reasons existed to make such an overdue study beyond the existence of a fascinating geophysical puzzle: the most important practical reason was concern about planning critical facilities such as nuclear and other power plants and determining whether nearby population centers needed to revise their building codes.

Johnston had to return to seismological square one. The plan was to map all the known midplate quakes, going back well beyond the instrumental record, which began a little before 1900, and into historical records. In North America this meant going back to the sixteenth century, in Europe some one thousand years, and in China two thousand years. Did such quakes occur only in certain kinds of stable crust — for example where old rifts existed as at New Madrid? Far more complex than that, their size needed to be calculated.

In this latter requirement, Johnston and several colleagues from the Geological Survey and the Center for Earthquake Research and Information at the University of Memphis were fortunate to have the excellent records kept by the British Raj of the horrendous earthquake in India's Rann of Kutch in 1819. They benefited also from the account of the quake and its aftermath in none other than Charles Lyell's *Principles of Geology*. A rann is a salt flat, and the Rann of Kutch, which flooded periodically, is located on the western coast of India and runs west-east. The 1819 quake there was an oddball among stable-crust quakes, and a helpful oddball. Practically every stable-crust quake occurs at great depths, buried under layer after layer of sediment, and they leave few traces on the surface of the earth. But the rupture at Kutch reached the surface and created a dramatic scarp almost ten yards high and more than fifty miles long. The locals called it the Allah Bund, which means the Wall of God. The scarp was the result of land to its north rising, and the land south of it (the rann) sinking below sea level. An imperial installation on the rann, Sindree Fort, sank to the point that the British soldiers escaped only by boat from the fort's uppermost turret.

The existence of the scarp gave Johnston and company exactly what they needed: an important step in the calculation of the Kutch earthquake's moment magnitude. This more precise measurement is derived from the actual events of the quake. From British records they had the length of the fault and the amount of slip, and with reasonable estimates of other matters — the depth of the rupture and the rigidity of the rock — they came up with the seismic moment magnitude (**M**) for the Kutch quake of 7.8. Equally important, they could use this estimate as a benchmark to test their magnitude estimates for other quakes that were derived from historical reports of intensity, in much the same method that Otto Nuttli had previously used to calculate the magnitude of the New Madrid quakes.

It turned out that the Kutch quake was exceeded in these stable-crust sweepstakes only by the three New Madrid quakes and the largest of the three Lisbon quakes. The largest of the three Lisbon quakes, which shook not just the earth but European philosophy, was a Magnitude 9.1. The New Madrid magnitudes, as estimated by Johnston and company, were December 1811 — 8.2, January 1812 — 8.1, and February — 8.3. It turned out that North America has been the most ferociously treated by such quakes. Six of the most powerful fifteen stable-crust quakes have occurred in North America, the others being one that struck Baffin Bay in 1933 (Magnitude 7.7), the largest of the three 1886 Charleston, South Carolina, quakes (Magnitude 7.6), and a 7.4 quake that struck the Grand Banks in 1929.

By way of comparison, the most powerful known quake of any type to hit anywhere in the world was the 9.5 quake that hit Chile in 1960, which Johnston estimated to have about the same amount of energy as a typical hurricane, but a hurricane's energy is released over a week or two, not two minutes. On the other hand, given the configuration of the ground of the eastern United States, the New Madrid quakes were felt over a greater area than any other known quakes.

In all, the Johnston study catalogued more than eight hundred stable continental crust earthquakes of Magnitude 4.5 or more, announcing their results in technical publications and then in *Scientific American* in 1990. The investigators assumed that there were many more particularly low-magnitude quakes that simply went unnoticed over the millennia. More than eight hundred 4.5 or higher quakes occur each *year* along plate boundaries like the west coast of this hemisphere, so these stable-crust quakes are vanishingly infrequent. Yet they are among the most potentially lethal.

More important than coming up with the Top Fifteen list, however, was discovering that these quakes did not occur at random. Some occurred at the site of rifts, places where continental crust had begun much earlier to break apart but had stopped doing so. But just as many occurred in what are called passive margins — places like Charleston, South Carolina, or Lisbon. Passive margins refer to areas that were created when rifts like the Mid-Atlantic Ridge actively began shoving the North American plate westward (and the European plate eastward) to create the Atlantic Ocean. The hot magma rising there began to cool, but not until the layer of crust it was forming was pulled and shoved, deforming and weakening in the process.

The stable-crust study went on from there to ask what made these weakened areas suddenly fail, and concluded that it was the buildup of pressure as the plates were shoved away from active rifts and into other plates moving in the opposite direction. Such pressures would be felt throughout the entire plate but have a disruptive effect where the plate is weak as a result of ancient tectonics — old faults like failed rifts or ancient mountain belts. And this seemed to be where the known stable-crust quakes have occurred — except in Australia which is an earthquake anomaly: the island continent is all stable crust and the scene of numerous earthquakes in at least the past six million years if not earlier, the largest of which have mostly occurred in ancient unrifted rock and have produced surface faulting. So, outside Australia, the stable-crust quakes all strike in rifted areas or passive margins, not at random, which is a tiny step toward forecasting such events.

On the other hand, forecasting stable-crust earthquakes is a hugely difficult problem — far more difficult than forecasting quakes on the margins of plates — never mind making pinpoint predictions. By far, most of these pesky quakes occur in places with deeply buried faults and they don't occur often. Indeed, of all the Magnitude 7 or larger quakes in Johnston's catalogue, none of them was a repetition of a previous event (taking the three New Madrid quakes as one event). If they are periodic, then the periods are very long ones, which makes sense, given the fact that the pressure on the weak parts of stable continental crust causes the surface to move a tiny fraction of an inch a year over thousands of miles. This amounts to infinitely slow deformation of the crust.

On the other hand, the deformation could be expected to concentrate in certain areas weakened by earlier tectonic events such as the Reelfoot Rift, and that might shorten the interval between quakes there. In the late 1980s, Johnston suggested that the only way to answer that question was from a then newly emerging subfield of earthquake science — paleoseismology. This is done not with a sophisticated array of high-priced seismographs, but with shovels and a backhoe. This task is, in many ways, easier in areas like the New Madrid Seismic Zone or near Charleston, South Carolina, where earthquakes tend to liquefy the soil and cause sandblows, sand craters, and sand dikes. These disruptions, eventually covered by sediments, will typically appear different in texture and color from the surrounding earth when a trench is dug through the area. At the time of their occurrence, they often have tree branches and other organic matter fall into them, and as the sandblow becomes covered with more soil, it and organic residues like tree bark are sealed into the earth. The organic remains can then be sent to laboratories specializing in radioactive dating techniques — in particular, carbon-14 dating — and the dates of the earlier quake can be ascertained. It seems also that early Indians often found the sandblows attractive places to set up camps — they were well-drained spots in an otherwise marshy world. Indian artifacts, particularly broken pieces of pottery, are useful in assigning dates to the earthquakes.

In the 1990s several workers from the Center for Earthquake Research and Information at the University of Memphis, along with several other universities, set out into the New Madrid Seismic Zone, and from forty or so sites they determined that there had indeed been previous quakes of comparable size to the quakes of 1811–1812.[2] They were dated to about A.D. 500, A.D. 900, and A.D. 1450. So it appears that the New Madrid Seismic Zone has periodically suffered huge quakes at least four times in the past 1,500-odd years, the recurrence intervals being roughly 350 years, 550 years, and 400 years. The average interval, then, is some 430 years. This is pretty thin data to go on for making any sort of forecast of when it might happen again, but there is a trick one can play on the data to make a pretty good guess.

Assume for the sake of argument that the occurrences of the earthquakes in this zone are at random times. The average interval, as noted, is about 430 years, which is the same thing as saying that the area suffered about 2.3 major earthquakes every millennium. This reduces the problem to a probability that statisticians can represent by points on a curve. Specifically, what we want to know is the probability of a monster quake happening again in the New Madrid Seismic Zone between now and the year, say, 2100. The answer is 20 percent.

That is about as precise as one can get about such a possibility, though some others have attempted greater precision in the prediction business — and two of these efforts, including one that rattled the New Madrid Seismic Zone and beyond, are the topics of the next chapter.

10

False Prophets

ONE CAN EASILY IMAGINE that the ranchers of Parkfield, California, were not about to lose a great deal of sleep over the prediction that, give or take four years and according to a bunch of university professors, an earthquake was scheduled to strike somewhere near the little hamlet where they picked up their mail and bought livestock feed in times of drought. Such a midsized quake was not likely to swallow up their herds or otherwise interrupt much of their lives. But let the word get loose of a date certain for a Big One in a crowded place like a city and the results can be devastating — even without the arrival of the predicted earthquake. Make that city one that is largely unprepared for a big earthquake — a city, say, in South America — and add two competitive units of the U.S. Department of the Interior plus an independent scientific watchdog group operating in an ambiguous scientific arena. Then throw in the State Department, the U.S. Agency for International Development, and its action arm, and season it all with a dose of journalistic enthusiasm, and you have the recipe for a political if not a geological disaster. This is exactly what happened when U.S. government scientist Brian T.

Brady derived, from laboratory tests on rock failure and what he called Einsteinian math, a "clock" by which the occurrence of earthquakes could be predicted virtually down to the hour.

All this came about in the 1970s, when the seismological community was bullish about how imminently they would be predicting earthquakes as a matter of near routine and when, through the U.S. Geological Survey, funding for seismological research was soaring. It was not surprising, therefore, that the government of Peru grew alarmed when Brian Brady, a research physicist in the U.S. Bureau of Mines, announced in 1976 that a Magnitude 8.4 quake would hit a few miles out to sea from Lima, Peru, in the late fall of 1980. Through his laboratory studies of rock failures in mining situations, he claimed to have developed a theory that yielded an earthquake clock of hour-to-hour precision. The quake he predicted for Peru would devastate the capital city, and what it didn't destroy would be done in by the ensuing tsunami: hundreds of thousands if not millions would perish. Indeed, it sounded as if Lima and the coastal cities and towns would see a gargantuan replay of an earthquake-tsunami combination that in 1746 had leveled the city. Since 1586, Peru had been struck by six major damaging quakes, with seventy thousand people dying in a 1970 quake in the northern part of the country.

The Brady announcement was taken seriously by the Peruvian government and its scientific community, which asked the United States for some corroboration that the prediction had merit. The chief of earthquake studies at the U.S. Geological Survey quickly branded it "far out" — it was, after all, based not on seismological fieldwork but on lab experiments (which, if true, would change seismology for good). At a later meeting of a panel of American earthquake scientists, however, Brady was grilled and the panel concluded that his theory enjoyed "a reasonable scientific basis for its validity."

Brady's theory included three series or types of precursors that were deterministic: in other words, if the area experienced the first set, the rest were inevitable. The rock "system" would march inexorably to "failure," meaning an earthquake. The same procession of events, Brady said, took place in a mine failure and a major earthquake. The scale of the events was not relevant but a matter of (complex, to be sure) mathematical adjustment. Brady had written in a scientific paper published in Europe that he had predicted an Idaho mine failure in 1975, a 1969 quake in the Soviet Union, a 1971 quake in the San Fernando Valley, an upstate New York quake in 1973, and two 1974 quakes near Lima, Peru. (In fact, except for the rock burst in Idaho, all these predictions were in fact "retrodictions," a word that geologists apply to retroactive predictions of earlier events, not an uncommon process in working out prediction schemes.)

The two 1974 quakes near Lima, he said, had been followed by an unusually short aftershock sequence and this had set the stage for the major quake that he forecast for the area.

Much of the seismic information about the 1974 quakes had been developed by Brady's colleague and friend, a geophysicist with the Geological Survey, William Spence, and the prediction of the coming quake came to be known as the "Brady-Spence prediction."[1] In fact, it was merely a "forecast" until 1977, when Brady specified the time, place, and magnitude, making it officially a prediction. A Magnitude 9.2 quake, he announced, would occur in late October or early November of 1981. He compared it to the 8.6 earthquake that had hammered Chile in 1960, then the largest to have occurred since the beginning of instrumental seismology. Back in Peru, the press got hold of the Brady-Spence prediction, but the (then) military government squashed any public notice of it lest panic occur. In 1978, a massive quake struck Mexico's coast, which made at least Peruvian earthquake scientists all the more concerned, and they asked again for confirmation from the American scientific community of the Brady-Spence prediction's accuracy. In reply, the USGS did not exactly impugn Brady's work but cautioned that earthquake prediction was nowhere near being a routine matter. "No one," the reply said, "yet has an accepted method or recipe for earthquake prediction." At this point, a new player entered the scene.

Within the Agency for International Development (AID) was a subagency, the Office of Foreign Disaster Assistance, one mission of which was to promote disaster mitigation by helping nations develop preparedness programs and plans for disasters such as earthquakes. The disaster office's science adviser, Paul Krumpe, would, in the months before the predicted quake, make this action-oriented agency a prime mover, keeping the Brady-Spence prediction alive and well and in front of the Peruvian government while a certain bureaucratic and scientific confusion turned into international controversy. The disaster office sought funding to help the Peruvian government prepare systems for dealing with such a catastrophe.

At this time in the United States, however, the prediction languished at lower bureaucratic levels: the secretary of the interior, for example, who is the boss of both the Geological Survey and the Bureau of Mines, had never heard of it. In Peru, not surprisingly, it was a major societal and political matter. (Members of the U.S. press covering Peru would say that they made nothing much of it themselves since, back home, the editors were interested in hearing only about political coups in Latin America.) Meanwhile, at those lower bureaucratic levels, the Geological Survey was showing signs of resenting the nonseismological upstart who came from a different agency than the one empowered to lead and perform seismological research. At the same time, people at the Bureau of Mines enjoyed seeing the survey "twitch and jump" in the face of the prediction.

In 1979, Brady changed his prediction, based on new information and his reworking of the math. First there would be a sequence of thirteen foreshock quakes over nine months beginning in September 1980, followed in July 1981 by a mainshock of 9.8, which would strike offshore from Lima and rupture the earth south into Chile, with an aftershock in April 1982 of 8.7, heading north toward Ecuador.

At this point, with a prediction of what would be the worst earthquake ever, many members of the Geological Survey began (mostly among themselves) to question Brady's methods, for example saying that a rock failure in a lab is simply different than the real world of an earthquake. At the same time, Brady reassured the Peruvian scientists that if the nine-month period of foreshocks occurred as predicted, then the rest would fall into place. Brady and Spence recommended that a series of monitoring devices and networks be deployed — seismometers, stress gauges, radon emission tests, and so forth, which they suggested could be financed by Paul Krumpe's office.

Events were spinning out of control. Something in the manner of Ross Stein's earthquake "conversations," the political tremors brought about by the Brady-Spence prediction would rattle around well beyond the initial site, Peru, and outside of the normal realms of seismological expertise. But unlike Omori's pattern of steadily diminishing aftershocks, these tremors only intensified. In July 1980, a new civilian government took office in Peru and, since the military regime had believed in the earthquake prediction, the new regime began by denouncing Brady as "an international terrorist." But with some persuasion by their own seismic scientists, they soon came to take him seriously enough to raise the restrictions on press coverage, permitting the prediction to become public knowledge in Peru.

The various U.S. agencies were now off on their own trajectories, with Peru asking the Office of Foreign Disaster Assistance for funds to support the creation of an evacuation plan, along with the installation of seismic equipment. The U.S. embassy in Lima, meanwhile, was trying to respond to the cries from Peru and maneuver through the turf-inspired interagency jealousies unleashed within the Department of the Interior. The Geological Survey was becoming ever more disenamored with the Brady-Spence equation (which continued to be revised by Brady to the point where Spence would finally withdraw his support). Soon Clarence Allen, the head of the National Earthquake Prediction Evaluation Council, newly organized by the Geological Survey, was alerted to the prediction and agreed that his council should review it.

By mid-autumn of 1980, the Peruvian president had taken a personal interest in the matter, the lead Peruvian scientist was pointing out that the time was getting short, the U.S. embassy in Lima was calling for help, and the United States had yet to come officially off the fence about the prediction. Then, in November, Clarence Allen, who was also a professor at the California Institute of Technology, was briefing the press, including reporters from the major dailies and one wire service, about earthquakes in general and provided a description of the Brady-Spence prediction. According to one observer, the journalists present "sat bolt upright and grabbed for their notebooks." Within days, major stories appeared around the country. For example, the *Miami Herald* shouted, "Great Quakes Predicted for Peru." The *New York Times* blared, "Huge Earthquake Predicted for Peru and Chile."

The Office of Foreign Disaster Assistance now took the lead, pointing out that even if the Brady-Spence prediction did not hold up, it was clear that the Andes were due for a big quake and that the United States should be providing technical and emergency planning aid to those who would certainly be affected one day soon. Interagency meetings convened; turf wars simmered. In Peru some complained that the worldwide attention that occurred once the American press had let the lid off Pandora's box would be likely to cause economic problems in Peru, where people were already booking flights out of the country around the predicted time of the quake. The U.S. embassy, getting into the lingo of things, reported that these "psychological foreshocks" were in the 2 to 3 range but would intensify, and complained of bureaucratic foot-dragging in Washington. This resulted, of course, in more interagency meetings. Some American scientists blamed the disaster office's Paul Krumpe for keeping the prediction alive and in the Peruvian forefront, but he and his superiors saw the prediction as a good excuse for beefing up not only Peruvian preparedness but the disaster office itself as well.

On January 26, 1981, Clarence Allen's National Earthquake Prediction Evaluation Council convened for its maiden meeting, charged with advising the Geological Survey on the scientific "validity" of earthquake predictions made by scientists either within or outside the government. What was to become essentially the trial of Brian Brady began — even as the entire matter had reached the assistant secretary level in the Departments of the Interior and State. Memos were flying back and forth suggesting two incompatible notions: that the evaluation council's hearings were too open to the press or that the council was a star chamber designed to discredit the prediction. Still others in government were fretting that if the council did not blow away the prediction, then what was Hawaii supposed to think, being right in the path of the tsunami associated with the prediction, which by now had been revised yet further upward to an earth-shattering 9.9.

As the "trial" opened, Brady spoke for a few hours explaining his theory and stating that his mathematics were based on Einstein's unified field theory. This meant little to the members of the council, who tried to halt the proceedings, stating that none of them understood such things and had not boned up on them. Brady was urged to move on to something the panel could understand. From there, things went from bad to worse, with Spence in the box apparently backing off from his support of the prediction, and then Brady, back in the box, still not providing a clear connection between his theory and the situation in Peru. Said one member of the panel (as the world press no doubt looked on with similar bafflement), "This is virtually incomprehensible. This sequence back and forth that we listened to this afternoon is almost impossible to follow, and there must be a better way to pull it together so that we can comprehend."

The next day, bewilderment merged into outright hostility, with Brady being accused at one point of deriving his mathematics from the Tibetan Book of the Dead, and berated for not providing the council with a straightforward mathematical equation that made sense to the members. In fact, here were two paradigms passing in the night, and finally, in response to another query, Brady said that if some large foreshocks did not begin by May in the vicinity of Lima, he would withdraw his prediction.

The evaluation council soon issued a statement rejecting the Brady-Spence prediction, saying that its decision by no means suggested that serious earthquakes would not occur in the region. Preparations for major quakes were appropriate. *Science* magazine, the organ of the American Association for the Advancement of Science, promptly reported all this under the headline "Prediction of Huge Peruvian Earthquake Quashed."

Even after the council's decision, the Office of Foreign Disaster Assistance kept in touch with Brady, asking him to speak about his prediction and about his theory, in which he continued to have faith. The days for the foreshocks came and went peacefully, though a few rumbles and the sudden appearance of a great number of fleas in Lima gave Krumpe reason to continue belief in Brady's prediction. The days for the great 9.9 quake came and went with no trouble. The Geological Survey had sent an official to spend doomsday in Lima to show its lack of concern. Frayed

184 | LOOKING BACK, LOOKING FORWARD

nerves in Peru began to repair themselves; the nation now was somewhat better prepared for and more knowledgeable about earthquakes, seismology, and emergency preparedness. Meanwhile most of the various U.S. scientific agencies and the evaluation council exonerated themselves, while knocking each other for one or another lapse in judgment. Mostly, Krumpe of the Office of Foreign Disaster Assistance got blamed for continuing to value Brady's prediction after it had been discredited. The disaster office even had the chutzpah to notify Secretary of State Alexander Haig about the controversy, remarking favorably on the Brady prediction well after it had been discredited.

No one came out of this affair smelling like a rose. Spence's career had been damaged despite his (late) apostasy. An embarrassed evaluation council needed to establish some protocols by which it would assess further earthquake predictions — rules that would permit it to seem less of a court-martial and more of a deliberative body. Brady continued to believe his theory was okay, only the math needed refining. But after doomsday passed without seismic incident, he gave up such refinements and returned to more standard Bureau of Mines safety research. The press was excoriated for so eagerly picking up on this prediction and uncritically blaring it far and wide, and while the press should not be exonerated for looking for a dramatic story as opposed to the facts, few members of the press are any better informed about science than virtually all the nonscientific agencies and the high-level bureaucrats in government. Nearly twenty years later a prediction for New Madrid would be treated just as ineptly.

The Browning Affair: Snake Oil and Hype

In 1990, for many people and especially for those living in or near the New Madrid Seismic Zone, what should have been confined to a few column inches on the inside pages of a supermarket tabloid or two became a serious matter. Many Americans learned for the first time that long ago the worst earthquakes ever to hit the contiguous United States had happened in this out-of-the-way place where earthquakes aren't supposed to happen, and New Madrid, after almost two centuries of being an utterly obscure backwater, was on the map again. On December 3, 1990, a handful of New Madrid residents rode into town on horseback and confronted some fifty or more television and print news crews from around the world who had been camping there for two days already. Down the street, in a show of what was either bravado or fatalism or both, but certainly of the broad kind of American humor, a local eatery known as Hap's Bar and Grill opened at 6:00 A.M. to host its "Quake, Rattle, and Roll Party."[2]

Just over a year before this fateful day of reckoning in New Madrid, the *Arkansas Democrat* carried a piece by the Associated Press entitled, "New Madrid Tremors Due, Forecast Says." This nondescript headline would have gathered virtually no attention — tremors happen almost every day in the New Madrid Seismic Zone, though most are below the radar of human sensitivity — except that a man named Iben Browning, who described himself as a climatologist and scientist, had said that conditions in the Seismic Zone were "ripe" for another huge earthquake.

Rather like a train leaving the station slowly but gathering speed as the moments go by until it becomes a blur of motion, the forecast, or "projection," as Browning called it, gained momentum over the months. Perhaps the most convincing part of Browning's claim was the additional claim that he had predicted — within hours — the Loma Prieta earthquake that struck San Francisco in the fall of 1989. This was the famous "World Series" quake, announced first to the world by sportscaster Al Michaels, who saw the outfield swaying in Candlestick Park and realized it was an earthquake. Browning explained that he had made his prediction on October 9, 1989, at a meeting of about 450 corporate executives at the Saint Francis Hotel in San Francisco just eight days before the earthquake. According to one of the executives there, Richard Howell, a spokesman for the Farm and Industrial Equipment Institute in Chicago, Browning predicted a major quake in northern California within three days of October 16. The Loma Prieta quake occurred on October 17. With such a credential, Iben Browning was a man to be taken seriously, and by many he was. As late as December 1, 1990, two days before the projected quake was supposed to devastate the New Madrid area, no less than the *New York Times* referred to Browning as both "a nationally known climatologist" and "a New Mexico scientist with strong scientific credentials in several fields, an author and inventor, though not a credited seismologist."

As the months went by after Browning's original projection, it gained the kind of specificity that would make it an official prediction. It would, Browning said, occur on or about December 3, 1990. It would be a high six or seven in magnitude, maybe bigger — in any event, enough to cause widespread havoc. It would occur because at that time a 179-year cycle would have reached the point where the moon, the earth, and the sun were all in a line, creating what sailors call high, high tides and exerting a massive force on the crust of the earth, particularly along the latitude upon which New Madrid sits — a place that had gone a very long time without a quake. The high tidal forces exerted on the earth can pull the crust upward in some places by as much as a foot, and this, Browning said, historically caused volcanic eruptions. Adding that earthquakes were just one of his sidelines, Browning reported that he was really a volcanologist. Be that as it may, Browning also gave New Madrid a fifty-fifty chance of seeing a major quake near the day specified.

Many stockbrokers and other gamblers, insurance actuaries, psychics, and even some scientists love cycles, be they about sunspots or years ending with zeros. All such are occasionally claimed to have physical significance in human affairs, and some do. Had journalists looked carefully, they could have found scientific papers such as one by Smithsonian volcanologist Tom Simkin suggesting that volcanic eruptions do coincide with the 179-year cycle Browning cited. But reporters could also have found more scientific papers pointing out that the same cycle does not always produce volcanic eruptions and that they also occur in clusters in noncycle years as well. Which is to say, the 179-year cycle is not predictive of volcanic eruptions — or, presumably, earthquakes.

Virtually all seismologists familiar with the New Madrid region saw the Browning prediction as baseless, but nonetheless, if asked, not one of them would answer categorically that a quake would not happen when Browning had predicted. Predicting that a quake will not happen at a certain time in a highly seismic zone is just as difficult a call as predicting that one will happen at a certain time. These same experts also concluded in the early months of the prediction that if they spoke out against it, it would simply give it more publicity: they assumed that it was sufficiently silly that it would not gain what today in show business is called "legs."

But legs are what the prediction did gain. In June 1990, six months before doomsday, both California's *San Jose Mercury News* and Arkansas's *Jonesboro Sun* repeated the Loma Prieta prediction claim and added that Browning also projected an earthquake in California's San Fernando Valley in 1971 that measured 6.5 on the Richter scale and killed seventy-five people, a 1972 killer earthquake in Nicaragua, and the Mount Saint Helens volcanic eruption. The press began feeding on itself. That same month a paper in Memphis added yet another successful

Browning prediction, that of the 1985 eruption of the volcano Nevada del Ruis in Colombia that killed as many as twenty thousand people. To this, in the same month, the *St. Louis Post-Dispatch* added a successful prediction of the 1985 earthquake in Mexico City. In July, the *San Francisco Chronicle* added the spectacular fact that Browning, in an updated Loma Prieta prediction a week before the calamity, had missed the exact time by a mere five minutes. How much of all this came from Browning himself has never been ascertained and probably never will be.

By now, people in important positions in emergency management were taking these news accounts seriously. In August, Missouri's *Springfield News-Leader* interviewed Candace Adams of the Springfield–Greene County Emergency Management Office. "I think everybody ought to take him seriously," she said. In August as well, Mississippi's *Jackson Clarion-Ledger* reported that the director of that state's emergency management agency had said that there "is a great deal of controversy about [Browning's] approach at arriving at these projections," but on the other hand "the fact that he had other projections on the money, you can't discount that." Indeed, throughout the region, even highly skeptical people in charge of emergency management were of the opinion that, given the widespread fear on the part of the populace, it would be irresponsible not to take Browning's prediction seriously enough to inaugurate various plans for evacuation and to launch various earthquake awareness programs.

Local seismologists, such as those at CERI, the earthquake program run by Arch Johnston at the University of Memphis, were beginning to speak out about the silliness of the Browning scenario, and the fact that it was based on a theory that had been utterly discredited as a cause of geological catastrophes, but by now the press was on a roll. The habit of most journalistic enterprises when something controversial comes up in an area where the journal has little knowledge, such as science, is to get the "two sides" to the story. It is presumed that there are always these two sides, and no editor wants to see his paper or TV program accused of being anything less than objective. Happily for this institutionalized tradition, there arose a seismological expert who wholeheartedly backed the Browning prediction, calling the man, among other things, a genius: "Perhaps the most intelligent man I have ever met." This was David Stewart, then a professor at Southeast Missouri State University in Cape Girardeau. In July 1990, Stewart began issuing bulletins to local governments in the region that they should definitely take precautions based on the prediction by Browning, who had such an astounding track record as a predictor of events. The press fastened on Stewart like lampreys on a mackerel, finding him an authoritative voice to counterbalance any story-spoiling naysayer. The journalists did not report that Stewart was the *only* seismologist to support fully the Browning prediction, while all the rest said, however tentatively, that it had no scientific merit.

By about this time, Iben Browning himself withdrew from the public eye, leaving any interviews up to either his much harried daughter or to Stewart, who appeared to revel in the attention. Browning was apparently quite ill at this time and died several months after his predicted doomsday, evidently from complications arising from diabetes. It should also be pointed out that a few days before doomsday in New Madrid, Browning, through his daughter, withdrew his New Madrid prediction, saying that the great quake would occur not there but most likely in Japan. That did not happen, and as everyone now knows, no earthquake occurred on December 3 or any time near that date. In fact, the New Madrid Seismic Zone, where hardly a day goes by without at least a seismic squiggle, was eerily calm for many weeks before and after Browning's predicted doomsday. Mother Earth was — at least locally — inert.

But by then the New Madrid "catastrophe" had engineered a

tremendous amount of activity — most of it useless. And a few oddball "coincidences" had heightened the activity, one example being that in the days preceding doomsday, a national TV channel aired a movie called *The Big One* that hyped the destructive nature of a big earthquake in a most terrifying collection of special effects and melodrama. Emergency preparedness officials in the surrounding states had dropped most of their duties to respond to plaintive calls from the public for information, and state emergency agencies spent about $200,000 on such responses, while their employees also attended seminars, tested sirens, set up emergency shelters with food and water, planned evacuation procedures and radio communication, and organized earthquake drills throughout the region. Firefighters and police in the thousands participated in drills, as did the National Guard of several states, including Missouri and Tennessee. In Arkansas, a drill presumed a Magnitude 7.6 quake with about five thousand fatalities and nearly a hundred thousand people homeless. Hotel reservations in regional cities collapsed. One estimate put the costs of dealing with the imminent quake at $200 million.

Less costly in any financial sense, churches preached quake behavior, one advertising its service with the following questions: "Preparing for the Big One? Are You Prepared for the Last One?" So-called survival revivals were held in Missouri towns, and a man in a van equipped with megaphones drove through New Madrid speaking of the end of the world.

Those who stayed on in New Madrid (not every one of its three-thousand-odd inhabitants did) seemed to take the imminent quake with the greatest calm, indeed with ingenious humor. Patrons of Hap's Bar and Grill, for example, nodded wisely over a series of numbers written on napkins: 1234567890. The 123 was December 3. The 456 represented 4:56 A.M., the moment when the alignment of the moon, earth, and sun would be strongest. The 78 represented Magnitude 7.8, and 90 was the year

1990. What more proof could anyone ask? The chamber of commerce sold T-shirts and other memorabilia.

The buildup was energized by stupendous media coverage. Browning himself was interviewed at one point on ABC's *Good Morning America*. A segment on the prediction appeared on NBC's *Today* show. Local television stations in the greater St. Louis area ran a constant barrage of information on the imminent quake and how to prepare for it. In the year leading up to doomsday there were at least 500 newspaper articles, and from June to December 1990, 343 articles appeared in the three newspapers local to the region. They came in a crescendo: in June there were 2, in July 25, in August 18, in September 22, in October 90, and in November 118. In December, with only three days of suspense left but some recapping to do later, there were 68 stories on the quake.

Southwestern Bell Telephone published an "Earthquake Guide," Wal-Mart supplied its customers with a brochure of "Helpful Hints to Prepare for an Earthquake," the American Red Cross issued its own pamphlet noting "27 Things to Help Survive an Earthquake," and the region's newspapers followed suit. In all, $22 million in earthquake insurance was sold (no doubt over the protests of the insurance companies that the prediction was questionable).

As the day grew nearer, schools in New Madrid and near neighborhoods closed down, as did many stores. A psychologist arrived in town and set up shop to help children deal with their presumably hostile feelings about the predicted disaster. According to witnesses, he suggested that they kick a puppet representing Iben Browning, and the children in turn simply stared at the man. Most of the fifteen hundred employees of an aluminum plant that was by far New Madrid's largest employer stayed home on December 3, and some businesses in Marked Tree, Arkansas, some ninety miles from New Madrid and the southern terminus of the New Madrid Seismic Zone, were closed. One couple left Marked Tree and escaped to Nashville, only to find upon their return that their house had burned down.

Reassessments

The question arises: Who were Iben Browning and David Stewart, who, between them, bamboozled a large segment of society of the most technologically and scientifically advanced nation on the planet in the year 1990, some five hundred years after the origins of modern scientific thought? While the information on the two men was readily available before as well as after the prediction gained its legs and ran roughshod, very few in the press looked into their backgrounds. One who did, Robert Bazell, the indefatigable NBC News science reporter, commented forcefully on the inconsistencies in Browning's prediction record, but his article appeared in *The New Republic* a week after doomsday fizzled.

Prior to his earthquake predictions, Iben Browning had become known particularly in business circles as a student of climate patterns and their effects on world food production, demographic changes, and socioeconomic matters, publishing a monthly newsletter from his home in Albuquerque, New Mexico, called the *Browning Newsletter*. He was also, from 1975 on, a paid consultant to numerous companies, the names of which he kept close to his chest. One of these, however, was Paine Webber's institutional sales division, whose senior vice president, John Coleman, said in 1988, "He's done a lot of good predicting. He's the most popular consultant we've had." Indeed, he was much sought after as a speaker at business meetings in the Midwest and on the West Coast.

As a climatologist, he was what is called an autodidact — that is, he was self-taught. He had no degree or other credentials in this field. He held a college degree in biology from the University of Texas, and soon afterward, in the sixties, he became enamored of the idea of controlling animals by planting electrodes in their brains. This was a time when a Spanish biophysicist named Delgado had astonished the world by being filmed in the costume of a matador as he faced an onrushing bull with sword and cape. The bull had electrodes planted in its skull, and just as it was about to slam into Delgado, he pressed a button that released an electric charge into the bull's limbic system, and it screeched to a stop. Browning took up the notion that whales, similarly hooked up, could be equipped with hydrogen bombs and directed — much as one directs a toy car with a radio setup — into Soviet ports to obliterate them. Evidently, Albuquerque's Sandia National Laboratories supported his research, being much in favor of such projects at the time. Using dolphins as spies or underwater suicide bombers was all the rage in those days. Sandia's interest faded soon, as did Browning's, who never did get to try his scheme on whales in the sea but instead only on goats in the desert, where the process failed.

By 1969, Browning had emerged as one of many consultants hired by the National Aeronautics and Space Administration to advise on the Apollo program, which was about to achieve the first manned landing on the moon. Browning advised NASA that the moon was so deeply covered with dust that neither a human-manned craft nor a robot could successfully land on it — in spite of the fact that several robotic landers had already dropped gently down onto the lunar surface and had sent photographs back that were published in virtually every newspaper in the country.

He moved on to climatology specifically and to making predictions in general, also co-writing a book entitled *AIDS* in 1988 and an earlier one, *Past and Future History: A Planner's Guide,* in 1981, in which he produced a nineteen-page list of "inferred events," meaning predictions of events that would take place between 1980 and 2010, among them the following typical ones:

> There is a fifty-fifty chance Egypt's Aswan Dam will be removed by atomic weapons.
> Arizona will become wetter than any time within the last one thousand years.
> Feudalism will sweep the earth.
> France will be deeply involved in retaking an empire.
> At thirty degrees south latitude in 1986 and at thirty degrees north latitude in 1990, earthquake and volcanic activity will be exceptional.
> "Quantitative" people will be sufficiently skilled to use robotic slaves, but "humanistic" people will turn to human slaves.

In this era, futurists were common figures on the scene and most of them offered more careful scenarios. Herman Kahn, for example, ran the Hudson Institute, a futurist think tank, and in 1967 he published a series of projections in *Natural History* drawn from his book *The Year 2000*. In addition to a graph showing the world's population for that year as six billion (high by a half billion) he adduced:

> Widespread use of computers in such activities as literature searches, medical diagnosis, traffic control, crime detection, and intellectual collaboration.
> New techniques for preserving the environment based on better ecological understanding.
> Development of new drugs and devices affecting emotions and thoughts to control memory, learning, relaxation, alertness, personality, and perceptions.

Cheap and widely available weapons systems and new methods for lethal chemical and biological warfare.

New kinds of cheap, convenient, and reliable birth control
techniques.

What of Browning's stunning prediction of the Loma Prieta quake?
It was not all it was cracked up to be, as it turns out. Rather late
in the crescendo of activity and nerves and fears (and, to be sure,
guffaws) leading up to the December 3, 1990, doomsday, the National Earthquake Prediction Evaluation Council realized that
ignoring the Browning prediction had not sent it off into obscu-
rity and ridicule but on the contrary had let it grow like some awful creature in a grade B science fiction movie. A panel was hastily convened consisting in part of seismologists familiar with the
New Madrid Seismic Zone, and they set about dealing with a
type of earthquake prediction that they had not, in fact, been
conceived to settle.

The council originated as a way of evaluating predictions
made by legitimate scientists — that is, seismologists — never
mind that its first task had turned out to be the Brady-Spence ca-
per that arose not from seismological studies but the physics of
mining. One of the first chores the council undertook in the
Browning affair was to find out about his now universally hailed
prediction of the Loma Prieta quake. They ordered and received
a transcript of his talk to the 450 business executives, including
Richard Howell, at the Saint Francis Hotel that fateful day a
week before the quake. What he actually said that day was simply: "There will probably be several quakes around the world
[of] Richter 6-plus, and there may be a volcano or two around
October 16."

He did not say where these quakes would occur, though most
of the audience presumably assumed that he meant nearby, the

meeting being in a highly active seismic zone where people were always wondering when the Big One would hit. At least most of the audience recalled after the fact that Browning had been specific. In any case, as a prediction, it was about as telling as predicting that the sun will rise the morning after a period of darkness. Quakes of Magnitude 6 or greater strike somewhere on earth on average about every three days. What's more, choosing New Madrid as a potential site for his next predicted quake was shrewd because none had occurred there recently, and giving it a fifty-fifty chance of happening was also shrewd. When it did not happen, he could say there was, after all, a fifty-fifty chance it would not occur.

In response to Browning's prediction of a new New Madrid catastrophe, the National Earthquake Prediction Evaluation Council's report did not appear until October 18, 1990 — one and a half months before the predicted quake — and the press did not exactly emblazon its findings across the front pages in banner headlines. The *New York Times* report on October 19 was buried inside the paper amid the advertisements: "Panel of Scientists Dismisses Warning of Missouri Quake." The story began by saying that the eleven scientists said "there was a long-term possibility of a major earthquake along the New Madrid Fault, but said there was no credibility in the widely circulated projection made by Browning." By then, the juggernaut of nerves, frets, and preparations was plowing along at too high a rate of speed to be stopped.

Subsequent research into Browning's predictions — those he claimed had turned out to be true — cast further doubt on the man. For example, his prediction that Mount Saint Helens would erupt was in fact made several days after federal officials had begun evacuating people from the area. His prediction that communism would cease to exist in its present form was made eight months *after* the collapse of the Berlin Wall. It is a bit scary that someone whose credentials never came even close to their claims of expertise could spend so long a time making a reputation and a living on the fringes of science. And it may be that seismologists have now learned to be quicker on the draw when it comes to evaluating earthquake predictions, even those made by people they consider out-and-out quacks.

As for David Stewart, he worked as a geology professor at Southeast Missouri State University and director of its Center for Earthquake Studies. Had reporters looked into his record, they would have perhaps taken his support of the Browning prediction with a little less enthusiasm. To be sure, Stewart was not just a seismologist but a local one who was deeply concerned that preparedness for a big earthquake was far too little in the region. Throughout the several months leading up to doomsday, his endorsements rang unequivocally in newspaper articles and local TV appearances. The memorandum he sent out to state and federal agencies ended with this strong statement: "It will be a tragedy if what Dr. Browning has forecast comes true on December 3. It could be a worse tragedy if it does not happen and people become cynical and unmotivated in earthquake preparation so that when the destructive quake does come they are not ready."

On October 21, two days after the National Earthquake Prediction Evaluation Council held its news conference, reporter William Allen of the *St. Louis Post-Dispatch* recounted Stewart's credentials. Stewart had been involved in a previous and, if anything, more bizarre earthquake prediction. In 1975, while teaching at the University of North Carolina at Chapel Hill, he grew concerned about an apparent bulge in the earth's crust accompanied by low-level seismic activity near a nuclear power plant under construction near Wilmington. Fearing an impending calamity, he advised the state's governor to stop the plant from going operational. The state thereupon requested a second opinion — from the U.S. Geological Survey. The survey's conclusion was that the bulge was an erroneous data point and the seismic records showed nothing unusual.

Undeterred, Stewart sought his own second opinion — from a woman named Clarissa Bernhardt of California. Bernhardt was a relatively famous psychic whose track record included predictions that Nelson Rockefeller would become president, Queen Elizabeth would resign, and John McEnroe would quit tennis and play baseball for the San Francisco Giants. Stewart invited Bernhardt to North Carolina and together they flew around the state in an airplane. At flight's end, Bernhardt predicted there would be a magnitude 8.0 earthquake near Wilmington, North Carolina, on or about January 17, 1976. Stewart vigorously backed up her claim.

January 17 passed without a hitch, and David Stewart was denied tenure at the University of North Carolina. He then went on — in an astonishing career move — to run a small publishing company devoted chiefly to publishing his own works on natural childbirth and midwifery. A decade later he somehow bounced back into seismology at Southeast Missouri State and met Iben Browning. Even though he could have claimed an out by pointing to the fifty-fifty nature of the Browning prediction, Stewart quit as director of the university's Center for Earthquake Studies, though he continued to teach there. Later, he and a colleague would produce some fairly sensationalized books about the New Madrid quakes of 1811–1812, in one of which he insists that, though the Browning prediction had not been fulfilled, one day when science has learned how to predict quakes with Browningesque precision, his theory of planetary influence will play a crucial role.

So what? it could be asked of this entire affair. There was no real or extensive damage. People got a bit of an education about earthquake preparedness and the causes (and noncauses) of earthquakes. And sure, some money was wasted. But the aftermath of this infamous nonevent saw the unleashing of a host of surveys and studies by seismologists and social scientists and even journalism professors to find out how the whole thing had gotten out of hand and how such matters should be dealt with in the future. Surveys suggested that many people claimed not to have been taken in by all the hysteria, but that their neighbors had. Other surveys suggested that many people, especially people new to the region, had learned of the 1811–1812 quakes and also how to prepare themselves for a large quake — a plus. But at the same time one of the other effects was to leave at least some of the public with a poor opinion not only of the media but of government and scierce as well. Such people might well not respond to another warning no matter how legitimate.

One researcher looking into the effects of the Browning affair found that many people really wanted to believe in the prediction, and would have even if the scientific community had made an earlier effort to discredit Browning's credentials and his prediction. These people, the researcher concluded, wanted to make up their own minds about such matters regardless of how technical and complex, and no matter what the preponderance of scientific opinion was.

In 1993, the U.S. Geological Survey issued its report on the Browning affair, replete with appendices of Browning and Stewart speeches and photocopies of hundreds of newspaper accounts. The report was pulled together by four scientists, two from the survey, one from St. Louis University, and one (Arch Johnston) from Memphis State University. One of the survey scientists was William Spence of the Brady-Spence affair, who had more direct experience with unorthodox earthquake predictions than anyone else in the field of seismology.

Among the survey's findings was the extent of the panic before December 3, 1990. Numerous earthquake-related institutions such as CERI at Memphis State and the Geological Survey were swamped with calls. They spent months "fighting unfounded rumors and rampant misconceptions about earthquake risks," with callers expressing fear, anxiety, hysteria, and panic. School closures caused about forty thousand students to miss school on December 3 in Missouri alone. The local Red Cross exhausted almost a quarter of a *million* copies of its earthquake hints.

What's more, many organizations had heightened the anxiety of the populace. Emergency preparedness officers (for obvious reasons) had taken precautions which suggested to the public that the prediction had merit. So did all the helpful hints issued by the Red Cross and Wal-Mart, none of which took the occasion to add a disclaimer about Browning's prediction. On the other hand the regional scientists were found to be relatively inexperienced in the business of dealing with quack predictions, and so was the press, not to mention the public.

Also, the disclaimers the scientists felt comfortable making could have been construed two ways — Browning's prediction was pseudoscience, therefore it was wrong, but scientists could not say an earthquake *wouldn't* happen around the predicted day, so Browning might be right. Not only that but there are many people who like to think there's something wonderful about mavericks who go up against the establishment.

The survey's report listed some of the lessons learned from this sad and tawdry affair. In-house press officers in scientific institutions need to alert the appropriate people in the institution when the media start hyping some potentially dangerous piece of pseudoscience. In the case of an earthquake prediction, it should be evaluated early on and the media need to be blitzed with coordinated packages of appropriate information, as do emergency preparedness officials. For example, the corroborating opinions of David Stewart played a huge role in validating Browning's prediction, and his opinion and credentials should have been loudly and promptly challenged by the scientific community. Meanwhile, those members of the press and television media who zeroed in on Stewart — the only seismologist in the region who gave Browning's prediction any credence whatsoever — were not being "objective" in their quest for *anyone* who would provide them with the fabled "two sides." What they accomplished, instead, was to throw an enormous and dangerous bias into their reporting.

Very much the same sort of thing can be seen these days when the handful of qualified scientists in the area of global warming who doubt that it is happening or are convinced it is not happening as a result of human endeavors are given equal weight by some journalists with the thousands who say it has already begun and is exacerbated hugely by humans burning fossil fuels.

11

New Madrid Redux

ARCH JOHNSTON IS A TALL, lanky man with a close-cropped beard, now white. Even when he is speaking about matters he knows as well as the topography of his hand, he speaks in slow, measured phrases, as if making absolutely certain that he is both right and explanatory. Since the early 1990s, Johnston had set himself the goal of finding out what had actually happened in the New Madrid quakes. Which quakes — the three big ones, the major aftershocks — had occurred on which faults, and what were their individual effects? He found himself poring over a lot of historical accounts, which he found somewhat ironic. His father had been a historian, and as a student Johnston did not think that much of such a study, science being where the real action was. But here he was doing a lot of historical as well as geological detective work. The more detail he could develop, the more reliably he would be able to pin down the actual magnitudes of the quakes.

In the course of his researches, carefully reading the eyewitness accounts, he was following the lead of Otto Nuttli, who had mapped the intensity of the December New Madrid quake, but

Johnston started with a more complete knowledge of where the faults were that underlay the New Madrid Seismic Zone. He found himself confirming some widely bruited details, and dismissing others as legend. Nuttli, in fact, had told him that he was unable to find any record of a church bell ringing in Boston on the night of the first quake, but it turns out that a Boston paper carried an item about a church bell ringing that night in Pennsylvania. From such small molehills can mythological mountains grow.

One of the greatest stories about the quakes was that the river — the mighty Mississippi! — ran backward for as much as three days. Of course that is fundamentally impossible, and many commentators have suggested that the notion of the river running backward was simply a slightly hysterical reaction to the turmoil of the water. Johnston tracked the Reelfoot Scarp south of the lake but also west to where it crossed the river (twice). As the scarp thrust upward in the vicinity of one of the river islands, it would have pushed the water backward (briefly), and that is exactly where an observer was when the quake hit, an observer who reported that the river ran backward.

Johnston was also able to demonstrate that contemporary accounts about the inundation of Little Prairie, leading to the citizenry's enforced exodus through the flood, actually conflated the great quake of December 16, which roused the townsfolk and sent them outdoors, with the huge aftershock that happened shortly after daybreak and brought about the inundation. But there was more to Johnston's efforts, of course, than setting the historical record straight. He wanted to pin down the moment to moment activity of the quakes, what faults they had run along, and other geophysical details.

Some seismologists were worrying that the faults that were lit up by seismographic recordings of the New Madrid Seismic Zone were not big enough in area to achieve magnitudes the size

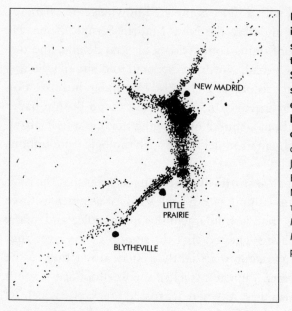

Figure 13. An artist's impression of the seismic signature of the New Madrid Seismic Zone. In a seismological map, each quake would be recorded as a tiny cross. Jake Page, derived from Arch C. Johnston, "The Enigma of the New Madrid Quakes of 1811–1812," *Annual Review of Earth and Planetary Science*, 24, pp. 339–384, 1996.

Johnston had assigned the quakes in the course of his global study of quakes in stable crust. But, for example, seismologists had found what might be another fault in the complex of faults over the Reelfoot Rift. It ran more or less northward and they called it the Bootheel Lineament. With more seismographic analysis, it later did turn out to be a fault capable of producing quakes.

Meanwhile, on the basis of various mathematical and other considerations, Johnston had scaled back the seismic moment magnitudes of two of the three major quakes: from 8.1 to 8.0 on January 23, and from 8.3 to 7.8 on February 7.[1] In all, a much clearer picture of the sequence of events had emerged, and Johnston published his new "model" in a technical publication called "The Enigma of the New Madrid Quakes of 1811–1812" in 1996, the same year that he issued his new magnitude estimates. Any such model is what might be called iffy, the less so the

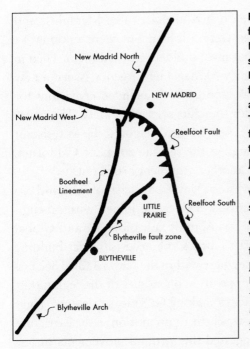

Figure 14. The seven known faults underlying the New Madrid Seismic Zone, in all some 220 miles of fault. The December quake ruptured the fault along the Blytheville Arch and the Bootheel Lineament. The huge aftershock that flooded Little Prairie struck on the Blytheville fault zone. The January quake struck the fault called New Madrid North, while the February quake struck the Reelfoot Fault and extended to the New Madrid West and Reelfoot South faults. Adapted from Arch C. Johnston, "The Enigma of the New Madrid Quakes of 1811–1812," *Annual Review of Earth and Planetary Science*, 24, pp. 339–384, 1996.

more it explains all the phenomena in a manner that fits with known science. For now, Johnston's model is the basic, accepted version of the titanic events of 1811–1812. What the ultimate cause was — what set these faults to slipping, what had caused the huge amount of pent-up energy to accumulate — remained an enigma. But with the model showing that a compact, deep network of faults could produce great quakes of around Magnitude 8, people generally accepted the new, slightly scaled-down magnitudes.[2]

What If

In 1991, a considerable argument broke out about building codes in Tennessee. Arch Johnston was among those who felt strongly that they should be strengthened in places like Memphis, and eventually stronger codes were adopted but were optional — which of course meant that most builders ignored them. Then in 2000 the Federal Emergency Management Agency issued a new national building code, IBC2000, which included provisions for establishing regional codes to achieve seismic safety. The codes called for buildings in the New Madrid region to be as resistant to earthquakes as buildings in the seismic zones of California. Perhaps not surprisingly, another argument broke out.

In this case, a Northwestern University seismologist and expert on plate boundary effects, Seth Stein, joined with an engineering consultant and a Los Alamos volcanologist and Global Positioning System expert to attack the need for strict building codes in Memphis and elsewhere within the New Madrid Seismic Zone.[3] They based their opposition on studies of the region they did by the accepted practice of making GPS measurements. Their studies showed that there was little or no motion of the earth taking place there. They suggested that ground movements of less than an inch a year mean it would take about a thousand years to build up enough strain to cause a Magnitude 7 earthquake. In one study they found no ground movement whatsoever. Thus, they said, the assigned magnitudes for the New Madrid quakes and any repeat performance were probably too high. They argued as well that the new code would add some $20 billion a year to Memphis's building costs while FEMA's estimate of potential earthquake damage was $17 billion.

They also pointed out that the very procedures needed to render a house more earthquake-proof would make it *less* safe in the event of a tornado, and that Memphis and much of the surrounding area are in Tornado Alley.

In essence, Stein was using concepts derived from such places as the San Andreas Fault to assess the likelihood of a new New Madrid quake. What causes quakes in California and along other plate edges is, as you now know, fairly well understood, but no one can say for sure what forces actually — ultimately — caused the great quakes of 1811 and 1812. Eugene (Buddy) Schweig, a member of the U.S. Geological Survey whose office is at the University of Memphis and who was deeply involved in the paleoseismological searches for signs of early quakes in the New Madrid Seismic Zone, took immediate issue with Stein. That motion has slowed, he said, or even stopped, does not mean that whatever really causes quakes in this region has died. The ultimate forces that cause these quakes may simply come and go. As for the tornado argument, Schweig, Johnston, and others point out that only relatively few buildings will be affected by a tornado roaring through on its path, but a big quake, and especially a great quake (Magnitude 8 or more) will threaten every building. Tornado damage is what experts call "acceptable risk." That means that the state does not insist on any special features in the building codes to protect against tornadoes. It is up to the individual. On the other hand what is unacceptable *social* risk is something that *is* defined in building codes. Schweig and others with a long association with the New Madrid Seismic Zone believe it is unwise to minimize the tremendous damage a reprise of the 1811 and 1812 quakes would cause.

If the building losses in Memphis added up to $17 billion, that would be just the beginning. There would surely be considerable loss of life there (and elsewhere), and other kinds of losses. Just what losses might occur if the New Madrid quakes happened one day soon has been looked into by numerous agencies, local and federal. The version that follows is based on various of these studies. But before looking at a scenario for the New Madrid region, it is instructive to look at the earthquake that struck Kobe in Japan on January 17, 1995. The Kobe quake caused the worst catastrophe in that nation since the history-changing explosions in Hiroshima and Nagasaki that ended World War II.[4] The city lies near the point where not two but three tectonic plates — the Asian, the North Pacific, and the Philippine — jostle for position. A quake was inevitable, and the Kobe quake, Magnitude 7.2, struck along a minor fault lying beneath the waters of Kobe harbor at about a quarter to six in the morning. The average vertical displacement along the fault was about one yard and the average horizontal displacement was almost a yard and a half, not enormous slippage, but sufficient to create a catastrophe. Japan is one of the most densely populated places on the planet, and Kobe is the nation's sixth largest city, with a population of 1.5 million. It is one of the largest container-ship facilities in the world.

In all, 190,000 houses and buildings were totally destroyed. Some 5,500 people perished and 317,000 were left homeless. Water and gas were cut off from nearly a million households, and about 150 fires raged for two days, further destroying residential areas in and around the city. Most of those damaged houses that remained standing were too dangerous to be repaired and had to be torn down. Two major expressways were severely damaged, and the trunk line of Japan's famous bullet train was out of business for three months. In one unforgettable scene, two large railroad cars were flipped up onto the roof of the train station like so many toys. Of the port city's 186 heavy shipping berths, 179 were inoperable, and the nation's exports dropped by 5 percent in the month after the quake even though the export trade was quickly shifted to other ports. The sum total of the direct and indirect costs — excluding loss of life and loss of income while the city spent three months reconstituting itself, repairing its utilities and infrastructure — was somewhere around $200 billion.

Now consider what would happen if New Madrid saw a repeat of the calamitous events of 1811 and 1812. Arch Johnston's study of midplate quakes and other assessments have shown evidence that such quakes as the New Madrid ones tend

Damage (millions of dollars)

State	Fire Damage	GROUND-MOTION DAMAGE Residential	Commercial	Total Damage
Alabama	291	1,152	956	2,399
Arkansas	2,646	8,978	23,726	35,350
Georgia	29	114	112	255
Illinois	853	3,427	3,875	8,155
Indiana	638	2,696	2,318	5,702
Iowa	1	4	4	9
Kansas	1	5	4	10
Kentucky	1,222	4,795	6,082	12,099
Louisiana	206	804	681	1,691
Maryland	0	1	1	2
Michigan	1	3	3	7
Minnesota	0	0	0	0
Mississippi	650	2,683	2,551	5,882
Missouri	1,581	5,854	9,806	17,241
Nebraska	0	0	0	0
New York	0	1	1	2
North Carolina	17	67	67	151
Ohio	462	1,744	1,593	3,799
Oklahoma	38	154	125	317
Pennsylvania	0	2	2	4
South Carolina	9	35	34	78
Tennessee	3,619	13,185	28,055	44,859
Texas	35	138	116	289
Virginia	4	17	16	37
West Virginia	8	30	28	66
Wisconsin	0	1	1	2
Total	**12,361**	**45,890**	**80,157**	**138,408**

Table 1. **Estimated building damage losses resulting from a recurrence of the December 16 earthquake.** From the Committee on Earthquake Engineering, *The Economic Consequences of a Catastrophic Earthquake* (National Academy Press, 1992).

to come in triplets.[5] A repeat sequence of the New Madrid quakes, the Federal Emergency Management Agency asserted, "would result in damages, disruptions, casualties and injuries on a scale never experienced from a natural hazard in the history of this nation." According to the National Academy of Engineering, whose Committee on Earthquake Engineering in 1992 estimated the building damage losses from a repeat of only the first of the three New Madrid quakes, the December 16 quake that achieved a magnitude of 8.2, twenty-four of the lower forty-eight states would suffer. Hardest hit, of course, would be Arkansas, Kentucky, Missouri, and Tennessee, with considerable damage to Illinois, Indiana, Mississippi, and Ohio. Yet even states as far away from the epicenter as Wisconsin, Maryland, and New York would not emerge totally unscathed.

The engineering study of 1992 came up with a total loss as a result of commercial and residential building damage from one quake of Magnitude 8.2 of about $140 billion. This included damage from both ground motion and subsequent fires. The estimate did not include other forms of direct economic losses that would occur. The huge region that would be affected is crisscrossed by electrical transmission lines, water and sewage lines, dams, highways and lesser roads, and railroads, as well as oil and gas pipelines — some of the nation's major movements of petroleum and petroleum products go through the New Madrid Seismic Zone or near to it. The region includes several nuclear power plants as well as many more conventional ones. Louisiana, where more than a billion dollars of residential and commercial structures would be lost, is also home to a major component of the Strategic Petroleum Reserve, located in salt cavities underground. Many of the region's rivers are dammed. Estimating which of these installations would sustain damage in an enormous quake is more of an art than a science, of course, but that many would be out of action for an indeterminate amount of time is a given. It took three months to repair the damage to the most critical installations in Kobe, such as the expressways. In the New Madrid Seismic Zone — and beyond — it would surely take longer. This is especially the case if the next New Madrid quake follows the pattern of the previous ones. From paleoseismological studies, it appears fairly certain that the predecessors to the 1811–1812 quakes came in fairly rapid-fire batches of three or at least two.

Consider (as FEMA has) what might be the fate of the largest city in the region — St. Louis. The number of casualties in the metropolitan area would depend to an extent on whether it was a work and school day. If so, most people would have left home for school, work, or shopping, with many country residents having gone into the city. But some 80 percent of the city's buildings — both residential and nonresidential — are made of unreinforced masonry, which is particularly vulnerable to earthquake damage. If the quake struck on a weekend or at night, most people would be home, and in the suburbs at least, most residences are predominantly wood-frame and less subject to the kind of damage that causes death and serious injury. Overall, casualties from a weekday quake might reach about forty-five hundred, while a weekend or nighttime quake would produce about four thousand, according to a 1990 estimate by the Federal Emergency Management Agency.

In either case, however, mayhem would ensue minutes after the quake. The metropolitan area's medical services would be swamped by injured people needing attention. Health care workers would find it difficult to respond to such emergencies because the damage to roads and highways would make it hard to reach the hospitals. Indeed, some of them would be among the dead and injured. Some of the hospitals would themselves be damaged severely enough to be out of business for several days if not weeks.

Flooding would be likely, which, along with other damage, would necessitate providing food and shelter for hundreds if not thousands of survivors. Relief efforts would be slowed, not to mention local, regional, and even national commerce, as the days went by. Emergency transportation requirements would have to

212 I LOOKING BACK, LOOKING FORWARD

be met chiefly by helicopter, for airport facilities would be limited
— because of either actual damage or loss of electrical power to
run navigational aids and runway lighting. The main utility systems — electricity, gas, water, sewer — might be unavailable for
days or even weeks to most residences, businesses, and other institutions. Port facilities along the Mississippi could be damaged
beyond use, preventing any assistance from arriving by water.
If the St. Louis arch still stood, it would preside not over the
proud Gateway to the West but over unprecedented suffering
and chaos.

Similar scenarios would almost surely be played out in other
great cities of the region — Memphis, Louisville, and Cincinnati,
to name three — and countless smaller towns and small cities
would be devastated. In 1811 and 1812, the three great quakes
caused gigantic waves in the Mississippi, waterfalls, and even retrograde motion of the waters. Whole sections of riverbanks collapsed into the river, causing mini-tsunamis. Today, from the
river town of Cairo, Illinois, south to Baton Rouge, Louisiana,
the levees that hold the great river to its present course would be
breached here and there, and the region would be severely
flooded. The floods would destroy not just homes and industrial
installations but farmlands as well. While in 1811, river traffic
was fairly common on the river, today it is greater by orders of
magnitude. Huge workboats pull, or guide, arrays of enormous
barges tied to one another, sometimes a quarter of a mile long.
Much of this traffic consists of toxic materials including much of
Chicago's sewage, and all this would spread across the flooded
landscape as the barges were lifted and dropped and capsized
amid ancient trees shooting up from the depths. Overall, such a
quake would have enormous industrial, agricultural, ecological,
and human ramifications. Fatalities could reach about 1 percent
of the twenty-two million people who live in the area of most severe damage — roughly some two hundred thousand individuals.

In addition to the direct losses noted above, indirect costs begin to pile up immediately. Economic growth would almost surely halt for several years at least. Property evaluations would drop. Huge losses would occur in personal and household income. Most of the private sector would be closed for an indeterminate amount of time owing to physical damage or to lack of supplies and other resources. The public sector — governments — would suffer losses of revenues as well as greatly increased expenses. Individuals and entire industries might well choose to relocate out of this sad and badly damaged region. The insurance industry would be obliged to sell off holdings in municipal bonds and other securities in order to meet their claims. Such a sell-off would cause a severe shock to the stock and bond markets. The interruption of vital supply lines in the region would cause the entire national economy to slow down for some period, several months at least. What would this all cost?

We have concluded that $1 trillion is not an unreasonable cost estimate. Such a huge cost would probably bankrupt some local and state governments and would put the federal government — already facing record debt — further in the hole for years if not decades to come.

Will it happen? Yes, in all likelihood, given what we now know about stable-crust earthquakes. When then?

The odds are that it won't occur in the lifetime of anyone living today. On the other hand, not knowing what exactly causes the network of faults in the New Madrid Seismic Zone to fail makes forecasting all the more difficult. But given the catastrophe a repeat of 1811–1812 would cause, discretion suggests that earthquake-proofing new buildings in the region be begun.

On August 16, 2003, a Magnitude 4 quake struck about four miles east-northeast of the town of West Plains, Missouri, which is about 120 miles west of New Madrid. Magnitude 4 quakes are not all that uncommon in the Seismic Zone and do not seem to worry people there too much. No harm was reported from the quake in August. Yet one can wonder if the quakes — mostly less than 4 — that rattle the area several times a year are aftershocks from nearly two hundred years ago, or just random seismicity . . . or foreshocks. Is a 4 and a handful of 2s and 3s each year enough to relieve whatever is causing the pressure, thus postponing the day when the Reelfoot Rift and the six others in the network fail and give the world another big one?

No one knows.

NOTES

INDEX

Notes

1. The World Gone Mad

1. The impressionistic description of events and all of the information and quotations in this chapter are to be found in the following volumes, of which Penick's history of the quakes is the most readily accessible. Penick, James Lal, Jr., 1981, 1994, *The New Madrid Earthquakes,* University of Missouri Press, Columbia. Audubon, Maria R., editor, 1897, *Audubon and His Journals, vol. II,* Charles Scribner's Sons, New York. Bringier, Louis, 1821, "Notices of the Geology, Mineralogy, Topography, Productions, and Aboriginal Inhabitants of the Regions Around the Mississippi and Its Confluent Waters," *American Journal of Science,* 3, pp. 15–46. Broadhead, Garland C., 1802, "The New Madrid Earthquake," *American Geologist,* 30, pp. 76–87. Flint, Timothy, 1932, *Recollections of the Last Ten Years,* edited with an introduction by C. Hartley Grattan, Alfred A. Knopf, New York. Fuller, Myron L., 1912, "The New Madrid Earthquake," *U.S. Geological Survey Bulletin,* 494. Haywood, John, 1823, *The Natural and Aboriginal History of Tennessee,* George Wilson, Nashville. Thwaites, Reuben G., editor, 1904, *Early Western Travels, vol. V,* Arthur H. Clark, Cleveland. This contains John Bradbury's account of his travels. Williams, W. D., "Louis Bringier and His Description of Arkansas in 1812," *Arkansas Historical Quarterly,* 48, pp. 108–136.

2. Dreams, Omens, and War

1. The characterization of this early attempt at town planning is from John W. Reps's article "New Madrid on the Mississippi," published by the Society of Architectural Historians, vol. 18, no. 1, pp. 21–26. Reps also supplies the quote from Victor Collot.

2. The details of Morgan's life come from Stanford historian Max Sevelle's biography with the overblown title *George Morgan, Colony Builder,* published in 1938. Sevelle did his best to celebrate the life of this interesting man, making as little as he humanly could of the fact that Morgan never was a colony builder. After the New Madrid flop, Morgan did have an important if brief role in American history. Former vice president Aaron Burr, after killing Alexander Hamilton in the famous duel, hatched a plot with General Wilkinson to persuade the western regions to secede from the Union. He mentioned his plan to Morgan on a trip through Pennsylvania, and Morgan promptly notified President Jefferson, who had Burr captured and tried for treason, the trial ending in an acquittal. Other references useful for this chapter were Henry M. Breckenridge's *Views of Louisiana: Together with a Journal of a Voyage up the Missouri River in 1811,* published in 1814 by Cramer, Spear and Eichbaum; Lewis Houck's *The First American Frontier: A History of Missouri from the Earliest Explorations and Settlements Until Admission of the State into the Union,* published in 1908 by R. R. Donnelly and Sons in Chicago. Yet another Missouri history of more recent vintage is Edwin C. MacReynolds's *Missouri: A History of the Crossroads State,* published in 1962 by Oklahoma University Press.

3. The subhead "The Entrails of the Times" was uttered in contempt by John C. Calhoun in denouncing those who believed in omens like comets. Accounts of the many superstitious beliefs invoked in the year 1811, along with the talk of war, are to be found in Penick's history, *The New Madrid Earthquakes.* For a fuller description of both the superstitions surrounding comets and also the Lisbon earthquake, the authors modestly suggest their earlier book, *Tales of the Earth.*

3. Pendulums and Polymaths

1. The Darwin quote is to be found in the *Encyclopaedia Britannica,* a much-overlooked source of rich material.

2. These early suggestions for causes of the New Madrid earthquakes are to be found in Penick.

3. The material on Jared Brooks is slim. He is discussed in Ben Casse-

day's 1852 volume, *The History of Louisville from Its Earliest Settlement till the Year 1852*, published by Hull and Brother of that city, and in Henry McMurtrie's *Sketches of Louisville and Its Environs*, published in 1819 by S. Penn, also of Louisville. The discussion of early seismographic instruments is from James Dewey and Perry Byerly's invaluable article "The Early History of Seismology (to 1900)," published in 1969 in the *Seismological Society of America Bulletin*, 59, pp. 183–227.

4. Mitchill's important contribution to what we know about the New Madrid quakes was published in the *Transactions of the Literary and Philosophical Society of New York*, no. 1, in 1814. The publication was based on a talk delivered in Washington, D.C., called "A Detailed Narrative of the Earthquakes Which Occurred on the 16th Day of December, 1811." A fine biography of this multi-talented man is *A Scientist in the Early Republic, Samuel Latham Mitchill, 1764–1831*, by Robert Courtney Hall, published by Columbia University Press in 1934.

5. The quote is from Susan Elizabeth Hough's book *Earthshaking Science*, published by Princeton University Press in 2002. It is as good an introduction for laymen to the realm of seismology as exists.

Part Two: The Earthquake Hunters

The epigraph is taken from Hans Zinsser's classic book about disease and human affairs, *Rats, Lice and History*, published in a reprint edition in 1984 by Little, Brown.

4. Myths, Maps, and Machines

1. This mythological lore is to be found in editor Maria Leach's *Funk & Wagnall's Standard Dictionary of Folklore, Mythology and Legend*, 1972.

2. For this chapter the Dewey and Byerly history was invaluable, as was a book compiled by the National Academy of Sciences and published by the Academy's press on-line in 2003: *Living on an Active Earth: Perspectives on Earthquake Science*. It contains a historical chapter called "The Rise of Earthquake Science." Far more technical is Benjamin F. Howell, Jr.'s *An Introduction to Seismological Research: History and Development*, published in 1990 by Cambridge University Press.

3. Lyell's visit to New Madrid is recounted in his book, *A Second Visit to the United States of North America*, published in two volumes by Harper and Brothers in 1849. The actual title of Humboldt's magnum opus is *Cos-*

mos: A Sketch of a Physical Description of the Universe. Volume I was published by Harper and Brothers in 1849 as well.

4. The only biography of Milne is by A. L. Herbert Gustas and P. A. Nott, *John Milne: Father of Modern Seismology.*

5. Finding Faults

1. Aside from specific reports about particular earthquakes, a good source of information, used throughout this book, is *Encyclopedia of Earthquakes and Volcanoes,* by David Ritchie and Alexander E. Gates, published by Facts on File in 2001.

2. Susan Hough's *Earthshaking Science* was helpful in this discussion. An invaluable reference here and throughout much of this book was Bruce A. Bolt's *Earthquakes,* an introductory textbook published by W. H. Freeman and Company, updated as recently as 1999.

6. Intensity, Magnitude, and Stars

1. Nuttli's 1973 paper on the intensity of the New Madrid quakes was called "The Mississippi Valley Earthquakes of 1811 and 1812: Intensities, Ground Motion and Magnitude" and appeared in *Seismological Society of America Bulletin,* 63, pp. 227–248.

2. Here again, Bruce Bolt's *Earthquakes* was useful.

7. Geophysical Leaps Forward

1. Abraham Oertel is profiled in the *Catholic Encyclopedia,* which is available on-line at www.newadvent.org.

2. Plate tectonics has been widely written about and at all levels. In addition to Wegener's own book, published first in 1914, one of the most important documents is the 1963 paper by Vine and Matthews, "Magnetic Anomalies over Oceanic Ridges," which appeared in *Nature,* 199, pp. 947–949. Other sources for this chapter include previously noted books by Bolt, Hough, and the National Academy of Sciences.

3. The Arkansas earthquake swarm was described by Arch C. Johnston in 1982 in *Eos,* 63, pp. 1209–1210.

8. Rifts, Plumes, and Reservoirs

1. Largely ignored for a very long time, the New Madrid quakes came in for a great deal of study beginning in the 1970s. The list of technical papers about the subject is by now a long one, and here is a sampling. Braile, Lawrence W., et al., 1986, "Tectonic Development of the New Madrid Rift Complex, Mississippi Embayment, North America," *Tectonophysics,* 131, pp. 1–12. Ginzburg A., et al., 1983, "Deep Structure of the Northern Mississippi Embayment," *American Association of Petroleum Geologists Bulletin,* 67, pp. 2031–2046. Johnston, Arch C., 1982, "A Major Earthquake Zone on the Mississippi," *Scientific American,* January, pp. 60–68. Johnston, Arch C., et al., editors, 1992, "The New Madrid Seismic Zone," *Seismological Research Letters,* 63, pp. 191–483. This reference contains twenty-five separate articles. McKeown, F. A., and L. C. Parkiser, editors, 1982, "Investigation of the New Madrid, Missouri, Earthquake Region," United States Geological Survey Professional Paper 1236. This reference contains twelve separate articles. Mooney, W. D., et al., 1983, "Crustal Structure of the Northern Mississippi Embayment as a Comparison with Other Continental Rift Zones," *Tectonophysics,* 94, pp. 327–348. Shedlock, Kaye M., and Arch C. Johnston, editors, 1994, "Investigations of the New Madrid Seismic Zone," United States Geological Survey Professional Paper 1538. This reference contains nineteen separate articles. Swanberg, C. A., et al., 1982, "Heat Flow in the Upper Mississippi Embayment," United States Geological Survey Professional Paper 1236-M.

9. The Art of Prediction

1. Johnston's study of midplate quakes was described in his and Lisa R. Kanter's article "Earthquakes in Stable Continental Crust," which appeared in the March 1990 issue of *Scientific American,* pp. 68–75.

2. A great deal of paleoseismic research in the New Madrid Seismic Zone has been carried out by Arch Johnston's colleagues, most notably Maritita Tuttle and Buddy Schweig. Some of the results are included in the papers listed here. Kelson, Keith I., et al., 1996, "Multiple Late Holocene Earthquakes Along the Reelfoot Fault, Central New Madrid Seismic Zone," *Journal of Geophysical Research,* 101, pp. 5161–6170. Tuttle, Maritita P., and Eugene S. Schweig, 1996, "Recognizing and Dating Prehistoric Liquefaction Features: Lessons Learned in the New Madrid Seismic Zone, Central United States," *Journal of Geophysical Research,* 101, pp. 6171–6178. Tuttle, Maritita P., et al., 1996, "Use of Archaeology to ———

Date Liquefaction Features and Seismic Events in the New Madrid Seismic Zone, Central United States," *Geoarchaeology,* 11, pp. 451–480.

10. False Prophets

1. The Brady-Spence prediction is the subject of a remarkable book by a political scientist, Richard Stuart Olson, with the help of a sociologist, Bruno Pedesta, and a professor of public affairs, Joanne N. Nigg. Titled *The Politics of Earthquake Predictions,* it was published in 1989 by Princeton University Press.

2. The Browning prediction received an enormous quantity of ink in the press, and an encyclopedia of more than a hundred newspaper articles and other pertinent documents is to be found in the Geological Survey's recapitulation of the event and attempt to learn whatever lessons arose from it. This report was by William Spence, Robert B. Herrmann, Arch C. Johnston, and Glen Reagor. More readable than one might expect from an official government document, and titled *Responses to Iben Browning's Prediction of a 1990 New Madrid, Missouri, Earthquake,* it appeared as United States Geological Survey Circular 1083. It remains the best single source on this affair, but other works were also consulted, including: Bazell, Robert, 1990, "A Little Shaky," *The New Republic,* October 19, p. A24. Browning, Iben, and Evelyn M. Gariss, 1981, *Past and Future History: A Planner's Guide,* Fraser Publishing, Burlington. Gori, Paula L., 1993, "The Social Dynamics of False Earthquake Prediction and the Response by the Public Sector," *Seismological Society of America Bulletin,* 83, pp. 963–980. Hubbell, Sue, 1991, "Earthquake Fever," *The New Yorker,* February 11, pp. 65–84. Shipman, Jill D., 1993, "The Impact of the Browning Prediction on Institutions," *International Journal of Mass Emergencies and Dangers,* 11, pp. 405–420. Wiener, Anthony, and Herman Kahn, 1967, "Excerpts from 'The Year 2000,'" *Natural History,* November, p. 10.

11. New Madrid Redux

1. Johnston's rationale for scaling back the New Madrid quake magnitudes is to be found in his 1996 paper "Seismic Moment Assessment of Earthquakes in Stable Continental Regions — III. New Madrid 1811–1812, Charleston 1886 and Lisbon 1755," *Geophysical Journal International,* 126, pp. 214–344.

2. The basic way that matters stand concerning the New Madrid quakes has been summarized in Arch Johnston's 1996 paper "The Enigma

of the New Madrid Quakes of 1811–1812," published in the *Annual Review of Earth and Planetary Science,* 24, pp. 339–384. Among other things, in it he traces each of the major quakes (and the huge aftershock after the December quake) to individual faults.

3. Seth Stein's attempt to diminish the threat of the New Madrid quakes appeared in *Eos,* vol. 84, no. 19, 2003, under the title "Should Memphis Build for California's Earthquakes?"

4. The Kobe quake was reported widely including in the following: Begley, Sharon, 1995, "Lessons of Kobe," *Newsweek,* January 30, pp. 24–29. Anonymous, 1995, *U.S. News & World Report,* January 30, pp. 38–44. "The Great Hanshin-Awaji (Kobe) Earthquake, January 17, 1995," published by the U.S. Department of Commerce, National Oceanographic and Atmospheric Administration, Boulder.

5. That quakes like the New Madrid quakes tend to come in triplets is to be found in a report, *Reassessing New Madrid,* by the participants in a 2000 New Madrid Source Workshop posted on the Web site of the Center for Earthquake Research and Information at the University of Memphis. That Johnston has long been arguing that more care should be given to building codes in the New Madrid area is to be seen in a report he and Robert M. Hamilton wrote in 1990, *Tecumseh's Prophecy: Preparing for the Next New Madrid Earthquake,* United States Geological Survey Circular 1066. Most of the estimating of damage and costs from a recurrence of the New Madrid quakes is the authors' own work, but they relied on various government reports, such as the Federal Emergency Management Agency's 1990 report *Estimated Future Earthquake Losses for St. Louis and County,* and the U.S. Geological Survey's 1885 report, edited by Margaret G. Hopper, *Estimation of Earthquake Effects Associated with Large Earthquakes in the New Madrid Seismic Zone.* Also useful was a report by the National Academy of Sciences' Committee on Earthquake Engineering, *The Economic Consequences of a Catastrophic Earthquake,* issued in 1992 by the National Academy Press.

Index

North America
  and continental drift, 117–19
  midplate seismic activity in, 132–
    33, 140–43, 169–75
  and Pacific Northwest seismic ac-
    tivity, 126–27, 152–54, 167
  See also United States; *names of
    specific states and countries*
North Anatolian Fault, 155
North Carolina, 12, 197–98
Northridge (Calif.), 162
Nuclear bombs, 111–14, 193, 207
Nuclear power plants, 137, 170,
  197–98, 210
Nuttall, Thomas, 43
Nuttli, Otto W., 100–103, 171,
  202–3

*Observations on the Geology of the
  United States* (Maclure), 42
Oceanography, 110–11, 118, 124
Oertel, Abraham (Oertelius), 116–
  17
Office of Foreign Disaster Assis-
  tance (U.S.), 179–84
Ohio River, 30, 34, 52
Old Faithful (Yellowstone geyser),
  151
Oldham, Richard D., 78, 80–81
Omori, Fusakichi, 76, 88, 98, 166
*Origin of Continents and Oceans,
  The* (Wegener), 115
Owen, Robert, 42
Owens Valley (Calif.), 84–86

P waves, 78–79, 81, 112
Pacific Northwest, 126–27, 152–
  54, 167
Pacific Ocean
  ridges and trenches in, 123, 125,
    128, 129

seismic activity around, 65, 125,
  126–27, 129, 132–33, 207–8
volcanism in, 144–46, 148
*See also continents and countries
  bordering*
Paine Webber company, 192
Paleoseismology, 85, 174, 207
Palmieri, Luigi, 70–71
Pangea, 121
Parkfield (Calif.), 163–65, 176
Passive margins, 172–73
Pelligrini, Antonio Snider, 117–18
Pendulums and springs
  Brooks', 44–48, 51, 68, 98
  Chinese, 45–46
  Drake's, 51–52
  Forbes', 69
  Milne and associates', 73, 74, 75,
    77
  Palmieri's, 70, 71
Penick, James Lal, Jr., 17, 38
Pennsylvania, 18, 26, 28, 48–49,
  203
Pennsylvania Gazette, 18
Perry, Alexis, 64
Peru, 176–84
Philadelphia (Pa.), 48–49
Pierce, William Leigh, 12–14, 34,
  56, 167
Pignataro, D., 47, 50
Plate tectonics. *See* Continental
  drift
Plumes (of magma), 146–49
Plutons (intrusions), 142, 143, 149
Poem upon the Lisbon Disaster
  (Voltaire), 32–33
Pompeii (Italy), 65
Power plants (and seismic activity),
  137, 156, 170, 197–98, 210
Predictions (of earthquakes), 34–
  36, 131, 133, 152–201